BAYESIAN STATISTICS FOR EVALUATION RESEARCH

CONTEMPORARY EVALUATION RESEARCH
A series of books on applied social science

Series Editors:
HOWARD E. FREEMAN, *Department of Sociology, UCLA*
RICHARD A. BERK, *Department of Sociology, University of California, Santa Barbara*

The CONTEMPORARY EVALUATION RESEARCH series meets the need for a monograph-length publication outlet for timely manuscripts on evaluation research. In the tradition of EVALUATION REVIEW (formerly EVALUATION QUARTERLY), studies from different disciplines and methodological perspectives will be included. The series will cover the full spectrum of substantive areas, including medical care, mental health, criminal justice, manpower, income security, education, and the environment. Manuscripts may report empirical results, methodological developments or review an existing literature.

Volume 1: ATTORNEYS AS ACTIVISTS: Evaluating the American Bar
Association's BASICS Program
by Ross F. Conner and C. Ronald Huff

Volume 2: AFTER THE CLEAN-UP: Long-Range Effects of Natural Disasters
by James D. Wright, Peter H. Rossi, Sonia R. Wright, and Eleanor Weber-Burdin

Volume 3: INEFFECTIVE JUSTICE: Evaluating the Preappeal Conference
by Jerry Goldman

Volume 4: REFORMING SCHOOLS: Problems in Program Implementation
and Evaluation
by Wendy Peter Abt and Jay Magidson

Volume 5: PROGRAM IMPLEMENTATION: The Organizational Context
by Mary-Ann Scheirer

Volume 6: RESEARCH DESIGN FOR PROGRAM EVALUATION:
The Regression-Discontinuity Approach
by William M.K. Trochim

Volume 7: DECISION ANALYSIS FOR PROGRAM EVALUATORS
by Gordon F. Pitz and Jack McKillip

Volume 8: BAYESIAN STATISTICS FOR EVALUATION RESEARCH:
An Introduction
by William E. Pollard

BAYESIAN STATISTICS FOR EVALUATION RESEARCH

An Introduction

WILLIAM E. POLLARD

CONTEMPORARY EVALUATION RESEARCH
A series of books on applied social science edited by
HOWARD E. FREEMAN and **RICHARD A. BERK**   8

 **SAGE** PUBLICATIONS Beverly Hills London New Delhi

TO MY WIFE AND MY PARENTS

For information address:

SAGE Publications, Inc.
275 South Beverly Drive
Beverly Hills, California 90212

SAGE Publications India Pvt. Ltd.
M-32 Market
Greater Kailash I
New Delhi 110 048 India

SAGE Publications Ltd
28 Banner Street
London EC1Y 8QE
England

Printed in the United States of America

Library of Congress Cataloging-in-Publication Data

Pollard, William E.
  Bayesian statistics for evaluation research.

  (Contemporary evaluation research; v. 8)
  Bibliography: p.
  1. Evaluation research (Social action programs)—
Statistical methods. 2. Social sciences—Research—
Statistical methods. 3. Bayesian statistical decision
theory. I. Title. II. Series.
HA29.P637 1985   361.6′072   85-14194
ISBN 0-8039-2509-3

FIRST PRINTING

Contents

Preface   7

1. Introduction   9
  What Is Bayesian Statistics?   12
  Bayesian Statistics and Evaluation Research   13
  The Development of the Bayesian Approach   17
  The Plan of the Book   21
  Further Reading   22

2. Overview of Classical Methods   23
  The Inference Problem   24
  Hypothesis Testing   25
  Interval Estimation   37
  Point Estimation   39
  Initial Precision Versus Final Precision
    of Experiments   40
  Further Reading   41
  Notes   42

3. Probability and Its Interpretation   43
  Axioms and Definitions   44
  Interpretations of Probability   46
  Degree of Belief and Probability   51
  Further Reading   57

4. Bayes' Theorem   59
  Bayes' Theorem for Events   60
  Bayes' Theorem for Random Variables   67
  Bayes' Theorem for Parameters and Observations   73
  Further Reading   76

5. Principles of Bayesian Inference   77

  Inference for a Normal Mean with Known Variance   78
  The Prior Distribution   87
  Comparison of Two Normal Means with
    Known Variances   102

A Note on Testing Point Null Hypotheses 118
Further Reading 121
Notes 122

6. Inference for Normal Means and Variances 123
The Model and the Likelihood Function 124
Comparison of Normal Means and Variances 146
Comparison of Normal Means with Variances Unequal 153
Comparison of Variances 156
Further Reading 157
Notes 158

7. Inference in Linear Regression 159
Simple Linear Regression 160
Analysis with an Informative Prior 168
Comparison of Group Means 172
A Note on Bayesian Inference for
Correlation Coefficients 174
Multiple Linear Regression 175
Comparing Regression Models 185
Further Reading 187
Notes 187

8. Inference for Binomial Proportions 188
Inference for a Single Proportion 189
Comparison of Proportions 207
Further Reading 207
Notes 208

9. Decision Analysis 209
Classical and Bayesian Decision Making 212
Decision Modeling 225
Further Reading 233
Notes 234

10. Summary and Perspective 235
The Major Advantages 235
Developments to Aid Application 237
Bayesian Diversity 238
Classical and Bayesian Methods in Perspective 239

References 241

Index 252

About the Author 256

There appears to be increasing interest among evaluators in Bayesian methods. One encounters frequent references in the evaluation literature to the Bayesian approach; yet, with a few exceptions, discussion is limited to brief comments suggesting that Bayesian statistics might be useful for certain problems in the evaluation of social and human service programs. Beyond this, however, many evaluators may not be aware of Bayesian statistics and what it has to offer. This is not surprising because much of the material on the Bayesian approach appears in sources outside of the evaluation and social research literature. Furthermore, Bayesian methods are seldom taught in social science statistics courses. Examination of widely used texts shows that Bayesian statistics is either not discussed at all or is relegated to an optional section that is seldom covered in courses because of time constraints. This is understandable to some extent because the Bayesian approach differs fundamentally from the conventional approach in a number of respects, and its consideration requires a significant shift in thinking about the nature of statistical inference.

This book is designed to provide an introduction to the logic of Bayesian analysis and to Bayesian methods for use in evaluation research. It addresses the basic questions that evaluators may have about the Bayesian approach: What is it? How is it different from other approaches? What advantages does it offer? How can it be applied? It is argued that Bayesian statistics has direct relevance to various types of problems encountered in evaluation research. Moreover, its relevance goes beyond the data-analytic kinds of issues that the term "statistics" calls to mind. In its most general form, Bayesian statistics includes a theory of learning from new data and a theory of decision making; this has implications for broader issues surrounding the use of evaluation findings. Indeed, Bayesian ideas have relevance in the fields of epistemology and moral theory. We do not pursue this here; however, persons interested in the philosophical analysis of the methods and objectives of evaluation might find this broad scope of Bayesian ideas intriguing.

This book is written for professionals and students, largely those in the social sciences, with an interest in data analysis for evaluation studies. It is assumed that the reader is familiar with the field of program evaluation as described in sources such as Fairweather and Tornatzky (1977), Morell (1979), Reiken and Boruch (1974), and Rossi and Freeman (1982), among others; the field of evaluation is not reviewed here. It is also assumed that the reader has a knowledge of statistics equivalent to the material covered in a graduate or advanced undergraduate course in social statistics. The level of mathematics required is essentially that of basic algebra. The treatment of multiple regression does go beyond basic statistics and algebra; however, the main points of the argument should be clear even without knowledge of multivariate distributions and matrix algebra. A review of certain statistical methods is included for comparative purposes. The references range from introductory- to advanced-level presentations.

The topic of empirical Bayes theory is, despite the name, distinct from the Bayesian theory to be covered in this book, and is not discussed. References to this area can be found in Susarla (1982); a fully Bayesian treatment is contained in Deely and Lindley (1981).

I would like to thank the editors of this series, Dr. Howard E. Freeman and Dr. Richard A. Berk, and an anonymous reviewer for their comments on an earlier draft of this material. The responsibility for the content of this book, however, lies entirely with me. I also wish to express my appreciation to Dr. Steven T. Levy, Chief of Service of Psychiatry at Grady Memorial Hospital; Dr. Dewitt C. Alfred, Jr., past Chief of Service; Dr. Jeffrey L. Houpt, Chairman of the Department of Psychiatry; and Dr. Bernard C. Holland, past Chairman, for providing the support that made completion of this project possible. I would like to thank the Emory University Computing Center for purchase and installation of some of the computer software used in this project. In addition, I want to thank Ms. Nancy Bowles for her excellent work in entering and formatting the manuscript on a word processor.

Portions of this material appeared in an earlier article (Pollard, 1983) concerned with Bayesian statistics and the utilization of evaluation research.

William E. Pollard
Atlanta, Georgia

1

Introduction

Evaluation research is generally advocated as a means of obtaining information to guide thinking and decision making concerning social and human service programs. Bayesian statistics specifies how new data should modify beliefs and affect decisions. That is to say, a theory and a technology exists for linking evaluation data and their use. The basic argument in this book is that Bayesian methods are therefore more directly related to the concerns of evaluators and users of evaluation findings than are the widely used conventional statistical methods. The purpose of this book is to provide a general introduction to the use of Bayesian statistics in evaluation research.

The methods that are commonly used to analyze evaluation data are, for the most part, those of what is called the *classical* school of statistics. The direction of this school was largely established by Ronald Fisher, and by Jerzy Neyman and Egon Pearson, in the first half of this century. Most of the ideas of hypothesis testing and estimation used by evaluation researchers are direct outgrowths of the work of these individuals. The influence of the classical school is described by Kyburg (1974: 22):

> By the nineteen-forties, the view, primarily associated with the name of Neyman, that the fundamental form of statistical inference was the choice between statistical hypotheses, had come to dominate other views in the English-speaking world. It merely dominated other views among statisticians; but it utterly overwhelmed other views among those whose interest in statistics was primarily practical. In medicine, in biology, in psychology, in sociology, in economics—everywhere that statistical hypotheses were formulated as such and tested directly—the vocabulary and formulations of the Neyman approach were so thoroughly ascendent, that for many scientists no other approach was even thinkable. "Significance level," "One-sided," "Two-sided," "Uniformly most powerful," "Power," "Error of type I," "Error of type II"—these terms were part of the vocabulary of scientific inference, and seemed to many who used them to admit of no alternatives. Even the Greek letters alpha and beta seemed to need no more explanation in statistics than pi in geometry.

Yet, among statisticians, there has been considerable debate over the foundations of statistics and the implications for statistical practice. One alternative to classical statistics that has figured significantly in this debate is Bayesian statistics. The Bayesian approach is not widely known among evaluators; however, it has a number of advantages over the classical approach for dealing with issues that arise in evaluation research. The purpose of this discussion is to make evaluators aware of this approach and its advantages.

There are three basic objectives in this presentation:

The first objective is to provide an overview of the logic of the Bayesian approach. Bayesian statistics is more than simply a collection of new or different formulas. It differs significantly from the classical approach at a very basic level. There are differences, for example, in the conception of probability, in the role of judgment in statistical analysis, and in the criteria for determining good statistical procedures. These differences are not simply theoretical issues of little relevance to the practitioner; they have direct implications for selection of methods of analysis and for interpretation of results. As a case in point, the procedures of hypothesis testing that occupy a central position in the application of classical statistics have a distinctly subsidiary position in Bayesian analyses. This review provides the justification for the specific Bayesian techniques to be discussed, and is necessary for understanding much of the debate between the advocates of classical and Bayesian methods.

The second objective is to provide a description of Bayesian techniques for data analysis and to illustrate their use. This provides an opportunity for a comparison of Bayesian and classical methods, and for discussion of the advantages of the Bayesian approach in evaluation research. Because most evaluators are familiar with classical methods, a comparative approach provides a useful way of highlighting important features and advantages of Bayesian statistics. The formulas and calculations are presented to illustrate how Bayesian methods could be used and what the results would be like, while making comparisons with classical methods. Bayesian results are expressed in the form of full probability distributions, and evaluating the results for various kinds of problems by hand can require a fairly extensive set of tables of distributions—tables showing only values for a few selected percentage points are seldom adequate for Bayesian use. Novick and Jackson (1974), for example, include close to 100 pages of tables. With the rapid increase in computerization of statistical procedures, researchers are less and less likely to conduct analyses by hand, and we analyze many of our examples, and evaluate the results, using computerized procedures. We do note a number of normal approximations, however, so that an interested reader could carry out analyses using the formulas that we discuss, and then refer the results to widely available tables of the normal distribution.

The third objective is to provide an introduction to the Bayesian literature. This book is an introduction to Bayesian statistics, and references are provided so the interested reader can further consider various issues and techniques. This literature is dispersed throughout publications in the fields of statistics, management, operations research, epidemiology, economics, philosophy, psychology, and sociology, and much of the Bayesian literature assumes a knowledge of the material referred to in the two previous objectives; the material in this book should make this literature more accessible. The reference list is not intended to be exhaustive. Houle (1978) describes a comprehensive Bayesian bibliography.

The general objective here is that of stimulating discussion, and application, of Bayesian methods. This could be of value for both the field of evaluation research and the field of Bayesian statistics. Although the focus of this book is on the former, the potential benefits to the latter deserve some mention. Although the origins of Bayesian statistics go back a couple of hundred years, only recently has it been a focus of substantial interest on the part of statisticians. Much of the work to date has been theoretical in nature and methods for application in many areas are still being worked out. Applications should provide a stimulus for the further development and refinement of Bayesian methods. Furthermore, certain areas in the social sciences are highly relevant to the application of Bayesian methods. As will be seen, the Bayesian approach makes salient a variety of subjective judgments that must be considered in any reasonable use of statistical methods. In fact, it calls for the quantification of certain judgments, and for their formal incorporation in the analysis. Social scientists have done a considerable amount of work on the measurement of subjective variables, and bringing this material to bear in applications of Bayesian statistics could be a substantial contribution to the field. Novick (1980) gives this point detailed consideration in a paper entitled "Statistics as Psychometrics."

It should be pointed out that consideration of the Bayesian argument has value beyond that of learning how to use Bayesian methods. Although much of the discussion in this book will treat classical and Bayesian methods as alternative methods of analysis, on a more general level the Bayesian approach can be viewed as an extension of the classical approach. Raiffa and Schlaifer (1961: 16) write that the Bayesian approach can be viewed as "a formal *completion* of the classical theory" in that it formalizes aspects of statistical analysis that are left to unaided judgment in the classical approach. Study of Bayesian methods often results in greater understanding of the classical theory. Leamer (1978) points out that the Bayesian view yields insights into the process of data analysis, and that one does not have to be a formal Bayesian to appreciate them. Consequently, confirmed users of classical methods may still find this presentation of value.

WHAT IS BAYESIAN STATISTICS?

Most of this book is devoted to answering this question, but by way of introduction, some general features may be highlighted. The focus of Bayesian statistics, in its most general form, is the revision of belief and choice of action in light of new data. The general approach incorporates both inference and decision making. Bayesian inference focuses on the modification of belief by new data. Decision making involves going beyond inference to make a choice among alternative courses of action on the basis of the new data. A distinguishing feature of the Bayesian approach discussed in this book is that beliefs about unknown quantities are expressed directly in terms of probabilities. These probabilities are commonly referred to as subjective probabilities. The revision of beliefs held prior to obtaining the data, to those posterior to the new data, is carried out through the use of a simple result in probability theory known as *Bayes' theorem.* The revised beliefs, expressed in terms of subjective probability, constitute the inferential result. Basically, inference involves making some appropriate probability statement about the phenomenon of interest. Decision making involves combining the probabilistic information with other information in order to choose among actions.

The Bayesian approach actually involves a formal theory of learning from the data along with a formal theory of decision making, and this is where it differs significantly from the classical approach. Bayesian statistics directly addresses the questions asked by the user of new data— How should these data modify my beliefs? How should they affect my plans and actions?—in a way that the conventional classical approach does not and cannot do, and the results, therefore, relate more directly to what the user wants to know. However, here, as anywhere else, there is no "free lunch." The Bayesian approach requires some hard thinking about one's inference and decision problems. Yet any reasonable use of the data requires such thinking, and the Bayesian approach makes this clear.

BAYESIAN STATISTICS AND EVALUATION RESEARCH

The argument here is that Bayesian statistics has direct relevance to a variety of problems that are encountered in the analysis and use of evaluation data. As we mentioned, evaluation research is commonly advocated as a means of obtaining information to guide thinking and decision making concerning social and human service programs. However, the path from the statistical analysis to the use of the results in thinking and decision making has always been somewhat vague, and classical methods have had little to say about what the user is actually supposed to think or do on the basis of the results. The Bayesian approach, on the other hand, explicitly deals with this issue. Moreover, as will be discussed in some detail, it is a normative approach—it spells out how data *should* affect one's thinking and choice of action.

The following are problems that have been mentioned in the evaluation literature. They range from what are seen as traditional statistical concerns to more general conceptual issues regarding the use of evaluation findings. In this book we consider how they can be resolved or clarified within the Bayesian framework.

CHOOSING AN APPROPRIATE
SIGNIFICANCE LEVEL

Virtually all inferential statistical analyses in evaluation studies involve tests of hypotheses. The choice of a significance level has an obvious impact on the result of the test and on the conclusion drawn. It is commonly believed that if the cost of error or loss function of the user of the findings were clearly specified, the selection of a significance level would be straightforward. An examination from the Bayesian perspective shows this to be insufficient. A reasoned choice of significance level must depend upon one's prior beliefs about the hypotheses under consideration—a consideration that lies outside the formal framework of classical inference. In practice, neither error costs nor prior beliefs are given much consideration in hypothesis testing; tests are carried out with certain conventional significance levels thought to be rigorous or stringent. Actually whether or not these conventional significance levels are really stringent depends again on prior beliefs about the hypotheses. In this book, we question the value of hypothesis testing in evaluation, and suggest Bayesian alternatives more suited to the problem.

STATISTICAL VERSUS SOCIAL SIGNIFICANCE

The use of arbitrary significance levels is the basis of this problem. The results of a test may be statistically significant but practically trivial; likewise, they may be nonsignificant, but of practical importance. This is a simple distinction, yet it leads to much misunderstanding. Even users of evaluation data with little or no statistical training "know" that results are supposed to be significant; consequently, trivial results can be touted as supporting one claim or another because they have been found to be "statistically significant." Likewise, one would certainly be subject to considerable criticism for advocating some action on the basis of results that were not found to be statistically significant. We argue that the use of hypothesis tests in many evaluation studies confuses a problem of inference with a problem of decision making. Again, we question the value of the classical test for making inferences or decisions about social programs, and present alternative Bayesian methods.

AFFIRMATION OF THE NULL HYPOTHESIS

The null and alternative hypotheses are treated asymmetrically in the classical theory of hypothesis testing, and the action implications of failure to reject the null hypothesis are unclear. Yet the affirmation of the null hypothesis may be of as much practical interest as the affirmation of the alternative hypothesis; indeed, in some instances a finding of no-difference may be a desired outcome, such as would be the case when there is an attempt to show that a new, less expensive treatment is just as effective as a more expensive treatment. In the Bayesian approach, the two hypotheses can be given equal status and there is no need to distinguish a null hypothesis and an alternative hypothesis. In fact, for inferences concerning treatment effects, there is little need to even deal with a special theory of hypothesis testing when one is using the Bayesian approach.

THE LONG RUN AND INTERPRETATION OF RESULTS

Significance levels and confidence coefficients refer to the performance of the method in the long run over repetitions of the same experiment rather than to the particular results that were obtained. They do not bear directly on the results of a single study. Yet users of research findings want to know what to think or do given the results at hand. In considering the results of a study, users often think in terms of the probability of a hypothesis being true or false or the probability that a parameter lies within some interval. Indeed, any reasonable use of the results requires this. However, classical significance levels and confidence coefficients do not directly relate to any such probabilities—this kind of thinking lies outside of the framework of classical statistics. The Bayesian approach addresses the user's questions more directly.

UNIQUE INFERENCE FROM THE DATA

The issue here is whether or not everyone should arrive at the same conclusion from a given set of data. Evaluation research is often seen as a way of resolving questions about social programs and it has been disconcerting to evaluators to find that even well-designed studies do not always lead different parties to the same conclusion. The classical theory has little to say about this interpretive phase of data analysis; the Bayesian theory makes clear that data do not necessarily lead to unique inferences. It delineates the conditions under which agreement will and will not be obtained.

COMMUNICATION OF UNCERTAINTY

The purpose of inferential statistics is to enable one to make inferences and decisions in the face of uncertainty. However, the classical approach does not provide a basis for researchers or users of research findings to express and communicate this uncertainty before or after the data are obtained. Expressions concerning how likely it is that program A performs better than program B on some measure of interest, for example, have no meaning within the classical approach. Indeed, expression of uncertainty is often obscured by disjunctive significant/nonsignificant outcomes of hypothesis tests because this is often interpreted as a yes or no answer to the question of whether or not an effect exists. Furthermore, everyday language terms such as "likely," "very likely," and "probably," are imprecise and have different meanings to different individuals. Bayesian statistics provides a precise and well-developed language—the language of probability—within which researchers and users can communicate and discuss the uncertainty involved in using evaluation findings.

RELEVANCE OF PRIOR INFORMATION

Evaluation studies are sometimes viewed as the only source of information concerning the evaluation issue at hand. However, users of evaluations are not suspended in some informationless state until the evaluation results come in. In some instances they may have considerable information about the program from other sources, and this prior information may be quite relevant to their conclusions and decisions. The Bayesian approach is a method for pooling prior and sample information; it specifies how this prior information should be revised by the new data. Furthermore, it spells out the conditions under which prior information will and will not have an appreciable effect on the inferences that are made.

USE OF EVALUATION DATA
IN DECISION MAKING

How should evaluation data be used in making decisions? Although evaluation is widely advocated on the basis of its potential use in decision making, just how evaluation findings should enter into the decision process is seldom spelled out. Bayesian decision theory specifies how the findings should be used in decision making. Furthermore, the decision framework makes clear the necessity of value judgments. The role of values in evaluation is, curiously, often misunderstood. Some discussions give the impression that decisions concerning programs should be made simply on the basis of "the facts" and that values should have no part in such decision making. The Bayesian framework shows how beliefs and values are both necessary components of decision making.

UTILIZATION OF EVALUATION RESEARCH

As I have pointed out, classical statistics has little to say about how the results should be used. It is left to the user to extract informally from significance levels and confidence levels the implications for what one should think or do. Bayesian statistics, on the other hand, explicitly deals with how the data should modify one's thinking and choice of action; it comprises a normative theory of research utilization.

These are examples of the kinds of issues into which the Bayesian approach can provide some insight. The reader is reminded again that there is no magic—any clarification is achieved at the cost of some hard thinking about issues that are either given only casual consideration or no consideration at all in the application of classical methods in evaluation. However, without such consideration, issues such as those discussed will remain to trouble evaluators and users alike.

THE DEVELOPMENT OF
THE BAYESIAN APPROACH

The purpose of this section is to provide some information on the origin of Bayesian ideas and their subsequent development. Knowledge of the history of quantitative techniques is seldom necessary for their application. However, some historical perspective is helpful in understanding the current status of Bayesian methods. There has been and continues to be considerable debate among statisticians regarding the foundations of statistics, and much of this debate centers on the differences between the classical and Bayesian schools. The stimulus for this debate is a paper written over 200 years ago by a Presbyterian minister, Reverend Thomas Bayes, entitled "An Essay Towards Solving a Problem in the Doctrine of Chances." Barnard (1958: 293), in a bibliographical note accompanying a reprinting of this paper, writes that it "must rank as one of the most famous memoirs in the history of science." Holland (1962: 451) notes that, in addition, it is also one of the least understood and controversial contributions. There has been much discussion of what Bayes "really meant" and, because it was published posthumously, whether Bayes actually intended to publish it in that form.

The increase in interest in the Bayesian approach over the last thirty years or so has led to better understanding of the ideas contained in the paper and their implications. Basically, the paper concerned the problem of inductive reasoning in statistics; that is, using data to make probability statements about hypotheses concerning the conditions giving rise to the data. Two aspects of this paper are important for the present discussion. The first is that Bayes' argument depended on use of a certain relationship involving conditional probabilities, which has become known as *Bayes' theorem*. Use of Bayes' theorem is fundamental to Bayesian statistics. The theorem can be easily derived from the axioms of probability theory and the definition of conditional probability, and by itself is not controversial. Differences arise, however, concerning the conditions under which it may be applied. The second aspect of importance is Bayes' use of the "principle of insufficient reason" or *Bayes' postulate*, as it has become known. The idea is that when we know nothing about the truth of the hypotheses prior to obtaining data, they can all be considered equally likely a priori. The Bayesian approach does not depend upon use of this postulate (a point that seems to have been overlooked in much of the early discussion of Bayes' paper); however, the notion of prior ignorance as a baseline or starting point for the accumulation of information has some intuitive appeal. A problem arises in that when strictly interpreted, this notion often leads to philosophical and mathematical difficulties. The expression of a state of ignorance in terms of probability is not as simple as it sounds. Indeed, the idea of "knowing nothing" is something of a contradiction in terms. De Finetti (1972: 159) remarks that the concept of "not knowing anything has never lead to anything other than confused discussion." As will be discussed later, Bayes' postulate and other expressions of the lack of prior knowledge are better viewed as an approximate representation of vague knowledge than as a strict representation of prior ignorance.

The ideas of Bayes were adopted by many, including the mathematician Laplace; however, in the mid to latter portion of the nineteenth century these views were heavily criticized, especially in England. This was due in part to the indiscriminate use of Bayes' postulate and in part to the development of an interpretation of probability that excluded degree of belief considerations. Although the Bayesian ideas continued to be of interest to philosophers, the founders of modern statistics focused on the development of methods that would appear to circumvent the need for Bayesian notions. From the late nineteenth century up until World II, the classical approach experienced tremendous growth and the Bayesian view was all but eclipsed. Actually, modern Bayesian theory (or neo-Bayesian theory, as it is sometimes called) had its origin in this period in the work of individuals such as Frank Plumpton Ramsey, Bruno De Finetti, and Harold Jeffreys; however, it did not have great impact at the time. A leading figure in the development of the classical approach was Ronald Fisher, whose work in experimental design, significance testing, and estimation established the direction of the field. He was highly critical of the Bayesian approach, and had a considerable influence in this respect.

Alternatives to Fisherian views on estimation and testing were put forth, within the classical framework, by Jerzy Neyman and Egon Pearson. Surprisingly, their work ultimately played a part in renewing interest in the Bayesian argument. Neyman and Pearson treated hypothesis testing as a formal decision problem. The decision typically involved a choice between a null and an alternative hypothesis, and considerations of cost of error were introduced in terms of the relative seriousness of Type I and Type II errors. This formulation was a major advance that cleared up a number of ambiguities in the theory of testing, and as the previous comment by Kyburg (1974) emphasizes, the Neyman-Pearson approach to hypothesis testing has become the predominant mode of statistical analysis used by researchers in applied fields today. In the 1940s, Abraham Wald extended these decision theory ideas in his *theory of statistical decision functions*. He went beyond the problem of deciding between hypotheses to develop a general method for making decisions in the face of uncertainty, and extended the formal analysis of cost considerations. His work had a significant impact at the time, and many statistical problems were reconceptualized as decision problems. Wald's theory involves the search for decision functions or rules specifying how one should act in light of any possible data that might be obtained in an experiment, in a way that would minimize loss.

Although Wald's work represents an extension of the classical approach, he did show that the decision rules having certain desirable properties were those that would also be arrived at by introducing Bayesian prior probabilities. Wald's (1950: 16) treatment of these Bayesian ideas, however, was primarily mathematical. He felt that prior probabilities would not be known for most problems, and that Bayesian solutions could not be obtained. As an alternative, he also advocated the use of mini-max procedures. However, for someone inclined to the Bayesian argument, the potential for further development along Bayesian lines was clearly there.

At the same time, growth of the decision theoretic formulation was stimulated by the publication in 1944 of the work of John von Neumann and Oskar Morgenstern on the theory of games. In game theory terms, statistical problems could be viewed as problems of decision making in games against nature. Furthermore, von Neumann and Morgenstern's formal justification of utility was to be very influential in subsequent developments in Bayesian decision theory.

In 1954, Leonard Jimmie Savage published *The Foundations of Statistics*, a book that was responsible for much of the renewed interest in the Bayesian approach. Drawing on the earlier work of De Finetti and Ramsey as well as von Neumann and Morgenstern, Savage was able to provide simultaneous justification for subjective probability and utility, and for the use of the maximization of expected utility as a criterion for decision making. Savage's argument was convincing to many and it did much to make subjective probability and Bayesian ideas "respectable." Lindley (1970: 37-38) notes that whereas the early Bayesians had to plead for the use of subjective probabilities, their existence was now established on the basis of a few mild assumptions. Interestingly enough the latter half of Savage's book was written to provide justification for classical methods from a subjective perspective. This portion of the book was not successful, as Savage notes in the preface to the second edition, and it is the first half of the book that has been influential.

In 1959, Robert Schlaifer published the first introductory Bayesian decision theory text; in 1961, Raiffa and Schlaifer published *Applied Statistical Decision Theory*, which remains a standard work on the theory of decision and on Bayesian distribution theory. Shortly after, Mosteller and Wallace (1963, 1964) published an application of Bayesian methods to resolving the question of authorship of certain of the Federalist papers, and this application gave increased visibility to the developing Bayesian approach.

From the 1960s on, the Bayesian approach has continued to develop. Yet much of this development has been accompanied by heated debate between Bayesian and classical statisticians. For persons interested in statistics as a mathematical system, the introduction into the formal theory of such seemingly vague elements as degrees of belief was anathema. Others worried that the introduction of these elements would compromise the objectivity of statistical methods. Still others expressed concern about measurement of subjective probability. We will touch on these issues in our discussion. More recently, the controversy surrounding the use of Bayesian methods seems to have died down some. (However, Hamaker [1977: 111] in a paper entitled "Bayesianism: A Threat to the Statistical Profession?" still expresses fear that adoption of the Bayesian approach "may seal the doom of applied statistics as a respectable and respected profession." The Bayesian responses to this are found in "A. Alleged Objectivity: A Threat to the Human Spirit?" by Good, 1978, and "The Mythical Threat of Bayesianism" by Moore, 1978.) Although the classical school still predominates, there has been growing acceptance of Bayesian methods and the number of publications in this area continues to increase. References to some of the more recent work will be provided in the discussion and readings.

THE PLAN OF THE BOOK

In Chapter 2, various elements of the classical approach are reviewed, especially those relating to the widely used methods of hypothesis testing. I assume that the reader has some familiarity with these topics, and include this review to emphasize certain aspects of classical logic. The use of classical inference procedures has become routinized to the point that investigators often simply scan computer output for p-values to see whether the results are significant or not, and the logic underlying the analysis is sometimes forgotten. Furthermore, in this chapter some notation is established and certain procedures are described for subsequent comparisons with Bayesian procedures.

In Chapter 3, some basic ideas in probability theory are reviewed, and different interpretations given to the notion of probability are discussed. It is at this basic level that differences between the classical and Bayesian approaches arise. These differences have important implications for the choice of methods of analysis and interpretation of results. Chapter 4 is devoted to Bayes' theorem. I show how it works with some simple examples and describe more complex forms necessary for making inferences about population or model parameters on the basis of observations. In Chapter 5, the normal model and inference for a normal mean is introduced, and a number of issues regarding the application of Bayesian methods and reporting of results are discussed. In addition, we consider the comparison of normal means, and contrast the classical and Bayesian approaches in some detail. In this chapter, and in following chapters, examples involving hypothetical data are considered.

In Chapters 6, 7, and 8, we discuss evaluation-related applications involving inference for normal means and variances, parameters of the normal regression process, and binomial proportions, and make comparisons with classical methods. Chapter 9 is an introductory discussion of the use of inferential results in decision making. We take a critical look at classical hypothesis testing as a decision procedure, and discuss an example of Bayesian decision making. Chapter 10 contains a summary and discussion of some general issues.

There is no standard system of Bayesian notation. The notation used in this book generally corresponds to the system used by Novick and Jackson (1974). There are two reasons for this. First, the discussion of Bayesian distribution theory and computational formulas closely follows that developed and presented in Novick and Jackson. Second, in the analysis of the examples, we use a package of computerized Bayesian procedures developed by Novick and his colleagues. Some of the input and output is displayed, and this is tied in with their system of notation. A more comprehensive, and more complex, system of notation is used by Raiffa and Schlaifer (1961).

One final comment on my approach is necessary. In Chapters 5 through 8, I present examples of Bayesian inference using hypothetical data. Bayesian inference involves the pooling of prior information with sample information to yield a result expressing posterior information. Because most evaluations are designed to be disseminated, and justified, to a wide and often diverse audience, the examples focus on analyses where prior information is either vague or is closely tied to previous data. Other kinds of prior information are not irrelevant, however, and can be incorporated in an individual's use of the data. In Chapter 5, we discuss the implications of the use of various kinds of prior information, and in Chapter 6, we discuss a computerized procedure for assessing prior belief and incorporating it into the analysis.

FURTHER READING

Arguments for the use of Bayesian methods in evaluation and policy research can be found in Crane (1982), Edwards et al. (1975), Fennessey (1972, 1976, 1977), Pollard (1983), and Wang et al. (1977). History of the development of the Bayesian approach can be found in Barnett (1982) and De Finetti (1972). The original reference for Bayes' theorem is Bayes (1958). Holland (1962) provides some biographical background for this work, and Seal (1978) provides a short discussion of Bayes' work. Fisher's response to Bayes, and its impact, is discussed in MacKenzie (1981) and Savage (1976). The context and significance of Savage's work is treated in Kruskal (1978), Lindley (1980), and in introductory comments to a collection of Savage's writings (American Statistical Association and The Institute of Mathematical Statistics, 1981).

2

Overview of Classical Methods

In this chapter, a review of some classical methods is provided. In order to emphasize the unique features of Bayesian methods, the discussion in subsequent chapters will involve comparisons of the classical and Bayesian approaches. The purpose of this review is to highlight certain aspects of the logic of classical methods and to provide a number of formulas for comparative purposes.

Classical statistical methods fall into two broad categories: hypothesis testing and estimation. Estimation, in turn, can be further subdivided into point and interval estimation. Actually, interval estimation is in some ways more closely related to testing than it is to point estimation, and this will be reflected in the order of discussion in this section. Testing has been the predominant approach to statistical analysis in evaluation research and in the social sciences in general; virtually all studies in which inferential statistics are used involve some test with a resulting statement about statistical significance or nonsignificance or, at the very least, a p-value. Accordingly, testing will be considered here in somewhat more detail than estimation.

THE INFERENCE PROBLEM

The basic problem of parametric statistical inference is to make statements about unknown quantities, or states of the world, that are not directly observable from observable random variables the behavior of which is influenced by the unknown quantities. These unknown quantities are viewed as parameters of a population or random process from which the observations are drawn or obtained; and the basis for inference is provided by a model of the population or process that expresses the probability of the observations conditional upon the unknown parameters.

The choice of a model is basic to both Bayesian and classical methods. For certain problems the nature of the data generation process may be well established on the basis of previous research, and choice of a model may be straightforward. For many problems, the form of the model is not established and the choice involves a judgment on the part of the investigator. This might involve extrapolation from other research, as well as considerations of parsimony and computational simplicity. Whatever the case, the investigator must make some judgment of a fairly specific nature about the form of the data generation process in order to get started.

Because the model involves a random component, the value of any particular observation is only one of many possible values that might be observed over repeated sampling from the population or process. The model probabilities are seen in the classical framework as expressing the relative frequency with which the potential observations would occur over repeated sampling. For a given model, with specified parameter values, the probability of any potential set of observations can be determined; the problem of inference, however, is to say something about the parameters given the observations rather than the other way around. The key to this is that the observations will generally not be equally probable for models with different parameter values, and this can be used in making inductive inferences about the parameters on the basis of the observations.

In most problems we do not work with the entire set of observations but rather with some statistic or statistics calculated from the observations. Yet the idea is the same—the value of the statistic that is obtained is only one of many possible values that could be obtained over repeated sampling, and the probabilities of the different values are expressed in a sampling distribution, the form of which is derived from the model. These probabilities are conditional upon a particular population model with particular parameter values. Typically, the statistic will not be equally probable in terms of sampling distributions conditional upon different parameter values, and this is used to make some inference about the value of the unknown parameter. Note that in this framework, the probabilities pertain to the observations and statistics, and not to the parameters. Although the parameters may be unknown, they are considered to be fixed—there is no basis for assigning a probability to a value of a parameter.

HYPOTHESIS TESTING

THE HYPOTHESES

In conducting a test, the investigator expresses the research hypotheses concerning the phenomena of interest in terms of statistical hypotheses, which are statements about the values of the unknown parameters. A hypothesis may specify exact values for the parameters of the model. This is known as a simple hypothesis; the probability model is completely specified by the hypothesis. For example, given a normal model with unknown mean μ and known variance σ^2, the hypothesis H: $\mu = 0$ completely specifies the probability model to be normal with a mean of zero and known variance σ^2. The hypothesis might also specify a range of values for the parameters, rather than a single point. This is known as a composite hypothesis and may be thought of as being made up of many simple hypotheses. Here the probability model is not completely specified. The composite hypothesis describes conditions that could be true of a number of models. For example, given the normal model just discussed, the hypothesis H: $\mu \leq 0$ describes a condition that would be true of any normal model with a mean of zero or less, and a known variance σ^2. Sometimes a simple null hypothesis is called a point null hypothesis, and this term is occasionally used to describe composite hypotheses involving two or more parameters in which the value of the parameter of primary interest is specified exactly. The hypothesis H: $\mu = 0$, involving the normal model with unknown variance, is a composite hypothesis because the value of σ^2 is unspecified; yet it is often referred to as a point hypothesis when μ is the focus of interest.

The investigator hopes to accept or reject a hypothesis on the basis of the observations. Usually, two hypotheses about the values of the parameters are stated: a null hypothesis (H_0), which is the hypothesis of primary interest in the sense that one hopes to be able to reject it, and an alternative hypothesis (H_1). The sampling distribution of some test statistic conditional upon the values of the parameters specified in the hypotheses provides the basis for accepting or rejecting the hypotheses.

HYPOTHESIS TESTING VERSUS
SIGNIFICANCE TESTING

It is important at this point to distinguish between two approaches to testing. The first approach, associated with R. A. Fisher, is known as *significance testing*. In significance testing, a null hypothesis is formulated that one hopes to disprove on the basis of the data. A test statistic is chosen for which extreme values are unlikely if the null hypothesis is indeed true. Given the sampling distribution of the test statistic conditional upon the value of the parameter specified in the null hypothesis, the consistency of the observed test statistic with what would be expected if the null hypothesis were true can be assessed. If the obtained or more extreme value of this statistic is highly unlikely in terms of this sampling distribution, it is considered to cast doubt on the truth of the null hypothesis. If the probability of the obtained or more extreme result is less than some significance level such as .01 or .05, the results are said to be significant at that level, and the null hypothesis is regarded as untenable. If the results are not significant, this does not mean that the null hypothesis can be accepted. Fisher (1973: 45) writes, "A test of significance contains no criterion for 'accepting' a hypothesis." The test allows one to reject, but not to accept, a hypothesis. In some instances, rather than make the dichotomous judgment of significant or nonsignificant, the investigator may simply report the *p-value* or the probability level at which the obtained result would be "just significant." It expresses the degree to which the data contradict the null hypothesis.

Two features of this approach deserve emphasis. First, only a null hypothesis is specified; no consideration is given to the consistency of the data with alternative hypotheses. This is part of the reason for not accepting the null hypothesis when it is not rejected—there may be other hypotheses that have not been considered that are more consistent with the data. Second, the probability expressed in a significance level or p-value is viewed as reflecting the strength of the evidence concerning the null hypothesis in that particular experiment. No specific procedures are provided for selecting a significance level. Presumably it relates to the cost of error, but this is not spelled out. Typically, small values are chosen.

The topic of testing was given a more systematic treatment by Jerzy Neyman and Egon Pearson in their theory of hypothesis testing. In their approach, an alternative hypothesis, as well as a null hypothesis, is stated, and the test is considered to be a procedure for making a decision between the two hypotheses in such a way as to minimize error. Acceptable levels of error depend upon the costs of error to the user of the test results. The focus of the Neyman-Pearson theory is on making optimal decisions, over the long run, about accepting or rejecting hypotheses rather than on making direct inferences about the validity of the hypotheses. Neyman and Pearson (1967a: 142) write, "Without hoping to know whether each separate hypothesis is true or false, we may search for rules to govern our behavior with regard to them, in following which we insure that, in the long run of experience, we shall not be too often wrong." Either hypothesis can be accepted or rejected.

By structuring the testing problem in this way, Neyman and Pearson were able to provide formal solutions to problems in the design of tests; in the theory of significance testing, on the other hand, problems such as choice of a test statistic were handled on an informal, intuitive basis. However, their formulation changed the problem somewhat; the emphasis shifted from inference concerning the valildity of hypotheses to the proportion of decisions that would be correct in the long run, if the hypotheses were true. Accepting and rejecting hypotheses have quite different meanings in the two approaches. Indeed, Neyman has referred to his approach as being concerned with inductive behavior in order to distinguish it from inductive inference concerning the validity of hypotheses.

The Neyman-Pearson approach is predominant today; nearly all presentations of testing for researchers are in hypothesis-testing terms. Yet researchers are often more concerned with inferring the validity of hypotheses than with making decisions entailing specific costs of error. Consequently, current practice is a somewhat inconsistent mixture of the two approaches—the mechanics of hypothesis testing are applied in carrying out a test, but the results are often interpreted in significance-testing terms. This has led to much confusion about what it means to accept and reject hypotheses, especially in the case of the null hypothesis; one occasionally encounters presentations of testing laid out in Neyman-Pearson terms accompanied by admonitions à la Fisher that hypotheses can only be rejected and never accepted. To be sure, there are reasons why a hypothesis tester might be reluctant to accept H_0 as a basis for action in certain circumstances, but the considerations are not the same as those of the significance tester. We will return to this issue later.

Even when the two approaches are clearly distinguished, there has been considerable debate over what the purpose of a test should be. Fisher felt that his methods were the appropriate ones for scientific research, whereas he considered the methods of Neyman and Pearson to be "statistics for shopkeepers" (Kyburg, 1974: 76). This issue is further complicated by the fact that testing has become something of an all-purpose tool in statistical analysis. Pratt (1965) lists nine different kinds of problems for which testing is usually applied, and questions whether such a general tool addresses any specific problem adequately. We will touch on this issue in more detail in subsequent chapters. This chapter will be primarily concerned with the Neyman-Pearson decision theoretic approach to testing, acknowledging that it is often used in ways that are not entirely consistent with the underlying theory.

THE DECISION

In order to make a choice between the hypotheses, the sample space comprising all possible samples of the observations is divided into two regions: one that is in some way consistent with H_0, and the other that is consistent with H_1. The test is basically an application of a decision rule of the form where, if the obtained sample falls into the region consistent with H_0, we accept H_0, and if it falls into the region consistent with H_1, we accept H_1. The null hypothesis is the hypothesis of primary interest and the language of testing reflects this. The two regions are usually labeled as regions of acceptance or rejection for H_0. It is understood that the region of rejection for H_0 is the region of acceptance for H_1 and vice versa.

The Probability of Error

The problem is to divide the sample space into regions of acceptance and rejection for the hypotheses in such a way as to minimize error.

There are two types of error possible here. The first type involves rejecting the null hypothesis when it is, in fact, true. This is a *Type I error* and its probability is designated by α. This probability is the probability of the observations falling into the region of rejection for H_0, when H_0 is true. If the null hypothesis is a simple hypothesis, α is the probability of erroneously rejecting this simple hypothesis. On the other hand, if the null hypothesis is a composite hypothesis containing more than one point, the probability of a Type I error may vary depending upon which point is the true value of the parameter(s) of interest. In this case, the maximum probability of a Type I error is designated by α.

The second type of error involves accepting the null hypothesis when it is false. This is a *Type II error* and its probability is designated by β. This is the probability of the observations falling into the region of acceptance of H_0 when H_1 is true. The alternative hypothesis is often a composite hypothesis, and the probability of a Type II error will generally vary depending upon which point in H_1 represents the true value of the parameter(s). The power of the test with respect to any specific point in the region included in the alternative hypothesis is $1 - \beta$. Power is the probability of rejecting the null hypothesis when the alternative hypothesis is true.

The Decision Rule

These two types of error are taken into account in choosing a decision rule. Ideally, one would like to use a test that would simultaneously minimize both types of error. This turns out to be impossible for the types of problems considered here. For a fixed sample size, a decrease in the probability of Type I error is accompanied by an increase in the probability of Type II error, and vice versa. This problem is usually handled by choosing an upper limit for the probability of a Type I error and selecting a decision rule that minimizes the probability of Type II error, or maximizes power throughout H_1, given this constraint.

For the tests to be described here, the theory specifies how the decision can be reduced to consideration of a test statistic that is calculated from the sample. The test is conducted by considering the observed value of the test statistic in relation to the sampling distribution of that statistic when the null hypothesis is true. If the observed value of the test statistic falls within a range of values called the *critical region*, the null hypothesis is rejected and the alternative is accepted. If, on the other hand, the observed value of the test statistic does not fall within this critical region, the null hypothesis is not rejected and is accepted.

In order to make this choice between the hypotheses, a decision rule is established in advance of the test that determines the critical region. Given an appropriate test statistic and the sampling distribution of the test statistic when H_0 is true, a decision rule is adopted where H_0 is rejected (and H_1 accepted) if the observed value of the test statistic exceeds some critical value, and H_0 is accepted (and H_1 rejected) when the observed value of the test statistic does not exceed the critical value. The critical value is chosen such that the probability of the observed value of the test statistic being greater than the critical value when H_0 is true, is less than some specified probability level. This probability defines the significance level of the test and sets an upper limit on α, the probability of Type I error. An observed value of the test statistic that exceeds the critical value is said to be significant. The choice of the significance level is left up to the investigator because it should reflect the cost of error to the investigator for a particular problem. Generally, it is thought that a Type I error is the more serious of the two errors, and that it is therefore considered appropriate to set a small upper limit on α, such as .10 or less. A detailed analysis of the seriousness of error lies outside the Neyman-Pearson theory of testing.

SOME TESTS ON NORMAL MEANS AND VARIANCES

Tests for a Mean

Suppose that some variable of interest X is normally distributed with mean μ and variance σ^2. We denote this as $X \sim N(\mu, \sigma^2)$. Suppose further that σ^2 is known and that we want to carry out a hypothesis test for μ. The hypotheses for a two-sided test are H_0: $\mu = \mu_0$ and H_1: $\mu \neq \mu_0$ where μ_0 is some null value. Here we are interested in whether the true value of μ is different from μ_0 in either direction—larger or smaller. The sampling distribution of $\bar{x}$, the mean of n independent observations on X, is such that the statistic

$$z = \frac{\bar{x} - \mu}{(n^{-1}\sigma^2)^{\frac{1}{2}}} \qquad [2.1]$$

has a standard normal distribution. The sampling distribution of this z statistic, conditional upon H_0 being true, is used in determining the critical region of the test. Specifically the test is carried out with the test statistic

$$\frac{x - \mu_0}{(n^{-1}\sigma^2)^{\frac{1}{2}}}$$

[2.2]

and the critical region is $z > z_{\alpha/2}$ and $z < -z_{\alpha/2}$. The critical value $\pm z_{\alpha/2}$ cuts off $100\,(\alpha/2)\,\%$ of the distribution of the test statistic in either tail. The probability that z will lie in either of these regions when H_0 is true is α. Hypotheses for a one-sided test are $H_0: \mu \leq \mu_0$ and $H_1: \mu > \mu_0$. Here we are concerned only with whether or not the true value of μ is greater than μ_0. The statistic 2.2 is the appropriate test statistic for this test also; however, the critical region for this one-sided test is $z > z_{\alpha}$. The critical value z_{α} cuts off $100\,\alpha\,\%$ of the distribution of the test statistic in the upper tail. The probability that the observed value of z will lie in the critical region if H_0 is true is, at most, α.[1]

In many problems the variance σ^2 will be unknown and the sampling distribution of the z statistic cannot be completely specified. Given the assumptions about the distribution of the observations, the statistic

$$t = \frac{\bar{x} - \mu}{[n^{-1}S^2/(n-1)]^{\frac{1}{2}}}$$

where

$$S^2 = \sum_{i=1}^{n} (x_i - \bar{x})^2$$

has a t distribution on $n - 1$ degrees of freedom. The sampling distribution of this statistic, when H_0 is true, is completely specified and t is the appropriate statistic for conducting one- and two-sided tests on the mean. (The classical unbiased estimator of σ^2 is $s^2 = S^2/(n-1)$ and s^2 appears in many classical presentations of statistics such as the t statistic just discussed. When the sum of squared deviations by itself is desired, it is often expressed as $(n-1)s^2$. The issue of unbiased estimation does not arise in Bayesian theory, and to facilitate comparisons with Bayesian material we use $S^2/(n-1)$ and S^2 rather than s^2 and $(n-1)s^2$ in our presentation.)

Comparison of Means

The logic of tests on the difference between two means is similar. Suppose that we have two variables of interest—$X_1 \sim N(\mu_1, \sigma_1^2)$ and $X_2 \sim N(\mu_2, \sigma_2^2)$—and would like to carry out a test for $\Delta = \mu_2 - \mu_1$. Consider, for example, the two-sided test with hypotheses H_0: $\Delta = \Delta_0$ and H_1: $\Delta \neq \Delta_0$ where Δ_0 is some null value. We obtain independent samples of n_1 and n_2 observations with sample means $\bar{x}_1$ and $\bar{x}_2$ from populations one and two, respectively. The sampling distribution of the statistic

$$z = \frac{(\bar{x}_2 - \bar{x}_1) - \Delta}{(n_1^{-1}\sigma_1^2 + n_2^{-1}\sigma_2^2)^{1/2}}$$

is standard normal. In the case where σ_1^2 and σ_2^2 are known, it is completely specified once we enter the null value Δ_0 for Δ, and it is used in determining the critical region of the test. A one-sided test could be carried out using this same sampling distribution, although the critical region would be different.

Likewise, when σ_1^2 and σ_2^2 are unknown but equal, the distribution of the statistic

$$t = \frac{(\bar{x}_2 - \bar{x}_1) - \Delta}{[(n_1^{-1} + n_2^{-1})(S_1^2 + S_2^2)/(n_1 + n_2 - 2)]^{1/2}}$$

which is standard t with $n_1 + n_2 - 2$ degrees of freedom, provides the basis for one- and two-sided tests.

Tests for a Variance

Tests on the unknown normal variance can be carried out in a similar manner. Suppose that we wish to test the hypotheses H_0: $\sigma^2 = \sigma_0^2$ and H_1: $\sigma^2 \neq \sigma_0^2$, where σ_0^2 is a null value. We obtain a sample of n observations with sum of squared deviations from the sample mean S^2. For the normal model with unknown mean, the test statistic $\chi^2 = S^2/\sigma^2$ has a chi-square distribution with $n - 1$ degrees of freedom. The distribution of this statistic when H_0 is true ($\sigma^2 = \sigma_0^2$) is used to determine the critical region of the test for both two-sided and one-sided tests concerning the value of the variance.

Comparison of Variances

Comparison of variances is usually carried out in terms of the ratio of the variances rather than their difference. Suppose that we obtain two independent samples of n_1 and n_2 observations with sum of squared deviations S_1^2 and S_2^2 from two populations with unknown means and unknown variances σ_1^2 and σ_2^2, respectively. Then we can make use of the fact that the statistic

$$F = k \; \frac{S_1^2/(n_1 - 1)}{S_2^2/(n_2 - 1)}$$

where $k = \sigma_2^2/\sigma_1^2$, has an F distribution with $n_1 - 1$ and $n_2 - 1$ degrees of freedom. Once we specify a null value k_0 for the ratio k, this sampling distribution is completely specified and provides the basis for one- and two-sided tests on the ratio of the variances.

In summary, these tests are all based on the known distribution of a test statistic conditional upon H_0 being true. Once a significance level is specified, the sampling distribution of the test statistic provides the basis for determining the critical region used in applying the decision rule.

THE ERROR CHARACTERISTIC OF A TEST AND CHOICE OF A DECISION RULE

The extent to which the investigator is protected against error when using a particular decision rule can be seen in the error characteristic of the test. Given a particular decision rule and the sampling distribution of the test statistic, an error characteristic curve can be plotted showing the conditional probability of making Type I and Type II errors for all possible values of the parameter of interest. Consider the one-sided test for the difference between two normal means $\Delta = \mu_2 - \mu_1$ where the variances are known and the hypotheses are $H_0: \Delta \leq \Delta_0$ and $H_1: \Delta > \Delta_0$. Curves for four decision rules with different significance levels are

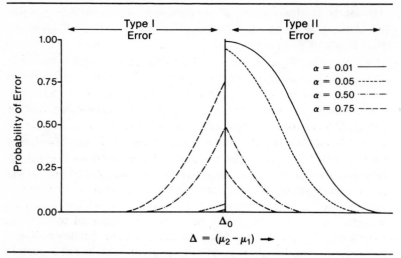

Figure 2.1 Error Characteristic Curves for 4 Decision Rules with Different Significance Levels

shown in Figure 2.1. For values of Δ less than Δ_0, we are concerned with Type I error; for values of Δ greater than Δ_0, the concern is with Type II error. It can be seen that increased protection against one type of error is achieved at the price of decreased protection against errors of the other type: Decision rules involving conventional significance levels such as .01 or .05 have the smallest probability of Type I error, but the largest probability of Type II error; the situation is just the reverse for decision rules involving the unconventional significance levels, such as .50 and .75.

In the Neyman and Pearson theory of testing, it is recommended that the choice of a decision rule be made by considering the error characteristic curves in light of the relative seriousness of the two types of error in the problem at hand. This is seen, however, as a consideration lying outside of the mathematical properties of tests, and the choice is left to the informal judgment of the investigator. Neyman and Pearson focused primarily on problems where Type I error is considered to be more serious than Type II error and many investigators thus routinely select a small conventional significance level in order to minimize the probability of Type I error.

Examination of the error characteristic curves also reveals some of the practical difficulty investigators experience in accepting the null hypothesis as a basis for action. It can be seen for a conventional decision rule with a significance level of .05, for example, that the probability of a Type I error is at most .05, no matter what the true value of Δ is. The probability of a Type II error, on the other hand, ranges from .95 to close to .00 depending upon the true value of Δ, which, of course, is unknown. This means that, for a range of values of the unknown difference that are greater than $Δ_0$, there is a large probability that the null hypothesis would not be rejected when, in fact, it should be rejected. Thus although the null hypothesis could technically be accepted with a known probability of error for any value of Δ as described by the error characteristic curve, the implications for subsequent action are unclear because the range of error probabilities is so great when Δ is unknown.[2] This reflects the basic asymmetry in the way that the two hypotheses are regarded in the classical test. Type I error is considered to be more serious than Type II error; and in order to limit the probability of Type I error, the probability of Type II error is allowed to be quite large in certain regions.

SOME COMMENTS ON HYPOTHESIS TESTING

There are three points that deserve emphasis here. First, rejection of the null hypothesis using a decision rule with a significance level of .05, for example, does not mean that the null hypothesis has a probability of .95 of being false, or a probability of .05 of being true. Within the classical framework, a hypothesis either is true or is not true—no probability is attached to the hypothesis indicating the degree to which it is likely to be true. The only probabilities involved in the test are probabilities of observations and sample statistics, given particular parameter values. One may informally think about the probability that a hypothesis is true; however, this lies entirely outside of the formal system of the classical test. The error probabilities refer instead to the proportion of errors that will be made in the long run using a specific decision rule. In other words, they relate to the method for choosing between hypotheses rather than to the hypotheses themselves. The error probability α is the probability of making a Type I error if H_0 is true; likewise the error probability β is the probability of making a Type II error if H_1 is true. These are conditional probabilities of error and not probabilities expressing the truth of the hypotheses. This is something less than what most users of research findings want to know and, as a consequence, these error probabilities are frequently misinterpreted as probabilities of hypotheses being true or false.

The second point is that although the test is clearly a procedure for making decisions between two hypotheses in light of the costs of error to the user of the findings, it is seldom explicitly used in this manner. As Silvey (1975: 161) writes, "The Neyman-Pearson theory provides an analysis of this problem which is to some extent half hearted. It recognizes that the two possible errors involved in a decision procedure, that is, in a test, are not equally serious and the criteria that it lays down for a 'good' test are based on this notion. But it does not involve any detailed analysis of the 'seriousness' of different errors." There is little discussion of assessment of loss functions and incorporating them in the analysis. Notions of loss are admitted only informally and those that are admitted are often of a very rudimentary nature. In many discussions of this issue, for example, it is implied that there is a fixed loss associated with each type of error. In other words, if we make an error in concluding that two treatments do not differ, this error is no more serious if they actually differ considerably than if they differ only slightly.

In practice, modeling of the user's loss function is seldom done. Indeed, many users of classical tests are not attempting to solve a specific decision problem. They simply want to get some idea of the tenability of the hypotheses given the data. If this is the case, then an accept/reject outcome using a preselected decision rule containing implicit notions of loss hardly seems appropriate. Of course, many investigators simply report p-values. Yet this seems out of place in the Neyman-Pearson theory because tests are constructed on the basis of maximizing power for fixed significance levels. This approach to testing as a decision tool is perhaps best suited for problems of acceptance sampling where, for example, lots of manufactured items are routinely sampled to determine the number of defective items before shipping to customers. Here a decision has to be made—to ship the items or not—and the costs of error can be specified in dollar figures. In addition, the "experiments" will be repeated over and over with new lots being continually manufactured, making the notion of long-run probability of error meaningful. However, tests are used for many other kinds of problems, and the extent to which they appropriately model these problems is open to question.

The third point is that a significant result means only that the observed value of the test statistic "signifies" that the null hypothesis should be rejected, in accordance with the decision rule established in advance of the test. It does not mean that the observed sample statistics are in some way significant by themselves, in the sense of being the true values of the underlying parameters. Nor does a significant result necessarily mean that the true underlying treatment effect is important, or of any practical significance. If the test is tailored to an individual's decision problem, then the significant result may have practical significance for that particular user. If other users have different decision problems, or no decision problem at all, then the results will not be of practical significance to them. The relationship between statistical and practical significance is further attenuated by the effects of varying sample sizes. Other factors held constant, the tests discussed here become more powerful throughout H_1 as the sample size increases. Consequently, we are increasingly likely to reject H_0 for values in H_1 that are close to the boundary with H_0. In other words, we are increasingly likely to reject H_0 when the true treatment effects are small. Rejection of H_0 with $\alpha = .05$, say, can have very different practical implications depending upon whether n = 10 or n = 10,000. Users of hypothesis testing results are occasionally warned that "statistical significance is not practical significance." Presumably, this means that users of the findings should make some adjustments or carry out reanalyses to bring statistical significance into line with what is practically significant for them.

INTERVAL ESTIMATION

The theory of interval estimation is closely related to that of hypothesis testing; it is sometimes described as an inversion of testing. The idea is to find an interval that will cover the true value of the parameter with some specified probability. Consider the two-sided interval estimate for a single normal mean where the variance is known. Because the statistic 2.1 has a standard normal distribution, we can find a number $z_{\alpha/2}$ such that

$$P(-z_{\alpha/2} \leqslant z \leqslant z_{\alpha/2}) = 1 - \alpha$$

Substituting the right-hand side of equation 2.1 for z, we obtain a two-sided interval such that

$$P[\bar{x} - z_{\alpha/2}(n^{-1}\sigma^2)^{1/2} \leqslant \mu \leqslant \bar{x} + z_{\alpha/2}(n^{-1}\sigma^2)^{1/2}] = 1 - \alpha$$

The probability $(1 - \alpha)$ is called the *confidence coefficient* of the interval. Similarly, it is possible to find a number z_α such that $P(-z_\alpha \leq z)$ equals $(1 - \alpha)$. A one-sided interval with a lower bound can then be obtained such that

$$P[\overline{x} - z_\alpha (n^{-1} \sigma^2)^{\frac{1}{2}} \leq \mu] = 1 - \alpha$$

Intervals for the difference between two means when the variance is known can be obtained in a similar manner. When the variance is unknown, the intervals for the single mean and the difference between two means are obtained by reference to the t distribution with the appropriate degrees of freedom. A criterion for constructing "good" intervals is that of minimizing the probability of false values of the unknown parameter being included in an interval, with fixed confidence level $(1 - \alpha)$. This parallels the criterion of maximizing the power of a test with fixed significance level α.

Interval estimation is sometimes advocated as an alternative to hypothesis testing. An interval estimate can be used to test many null hypotheses simultaneously because any null hypothesis involving a value of the parameter that lies outside of the interval would be rejected by a test of significance level α for that particular sample. In addition, the size of the interval provides some indication of uncertainty about the value of the unknown parameter. Of course, the error characteristic of the test also shows the uncertainty involved, but it is rarely examined. Furthermore, in some instances, it may be difficult to identify particular hypotheses of interest, and "bracketing" the true value of the parameter with an interval may be a more useful way of expressing the results.

The meaning of the confidence coefficient, or probability associated with the interval, requires some comment. Interval estimation procedures allow one to make statements about the value of a parameter such that, in the long run, $100 (1 - \alpha) \%$ of these statements will be true. The confidence coefficient $(1 - \alpha)$ associated with the interval does not mean that the true value of the parameter has a probability of $(1 - \alpha)$ of lying in the interval. In the classical framework the true value of the parameter either lies in the interval or it does not—no probability is attached to the value of the parameter. It is the interval that is random, and the probability relates not to the unknown parameter, but to the method for constructing the interval; in the long run, intervals constructed according to this method will contain the true value of the parameter $100 (1 - \alpha) \%$ of the time. Suppose that the confidence coefficient is .95. Then, before the sample is obtained, there is a 95% chance that the interval will cover the true value of the parameter. Any particular interval is simply one outcome of a random process that generates intervals such that, in the long run, 95% of these intervals will cover the true value of the parameter. Once the actual end points of the interval are calculated from the data, then the parameter is either in that interval or it is not. As Pratt (1961: 165) comments,

> Unfortunately we don't know which. We think, and would like to say, it "probably" does; we can invent something else to say, but nothing else to think. We can say to an experimenter, "A method yielding true statements with probability .95, when applied to your experiment, yields the statement that your treatment effect is between 17 and 29, but no conclusion is possible about how probable it is that your treatment effect is between 17 and 29." The experimenter, who is interested not in the method, but in the treatment and this particular confidence interval, would get cold comfort from that if he believed it.

Again, this is something less than what users of research findings want to know. The correct interpretation of a confidence interval is not intuitively appealing, and the confidence coefficient is usually misinterpreted as the probability that the parameter lies in the interval.

POINT ESTIMATION

In problems of point estimation, we want to choose a single number that will be a good estimate of a parameter of interest with respect to certain criteria. We cannot directly assess the quality of an estimate because the true value of the parameter is unknown. However, some assessment can be made by considering the performance of an estimation procedure over the long run, that is, by considering its performance in repeated samples. We will use the term *estimator* here to refer to the estimation procedure, and the term *estimate* to refer to the particular value that is obtained by applying the estimator to the observations.

A number of properties of estimators are considered to be desirable:

(1) Unbiasedness. The expected value of the estimator should equal the value of the parameter being estimated.
(2) Consistency. As sample size increases, the probability that the estimate is arbitrarily close to the value of the parameter should increase. In other

words, the estimate should tend toward the true value as the sample size gets larger.

(3) Efficiency. The variance of the sampling distribution of an estimator should be less than that of alternative estimators.

(4) Sufficiency. An estimator is sufficient if no other statistic from that sample provides additional information about the parameter. This means that the estimator contains all the information in the sample regarding the parameter.

There is no single method for constructing estimators that will be optimal in terms of all these properties. A variety of methods are in common use. Zellner (1971b) for example, lists some seventeen different methods that have been proposed for use in econometric estimation. Because there will seldom be one optimal estimator with respect to all criteria, subjective considerations concerning the relative importance of these criteria and issues such as the cost of data and computational difficulty may be required to make a choice among alternative estimators. As Cox and Hinkley (1974: 251) note, "any replacement of an uncertain quantity by a single value is bound to involve either some rather arbitrary choice or a precise specification of the purpose for which the single quantity is required." The commonly used classical estimators are not based on detailed specification of the problems involving their use, and the choice is therefore somewhat arbitrary. In certain instances, point estimation can be approached as a decision problem and a clear solution to the problem can be obtained. This requires, however, a detailed specification of the user's decision problem, including elements not formally considered in the classical framework.

Again, the quality of estimation is largely assessed in terms of the long-run performance of the estimators, or method of obtaining estimates, and does not relate directly to an estimate that we might obtain in a particular sample. The obtained estimate is simply one outcome of a random sampling process, and may or may not be close to the true value of the parameter.

INITIAL PRECISION VERSUS
FINAL PRECISION OF EXPERIMENTS

We have emphasized that in all areas, the only probabilities involved are the probabilities of observations and sample statistics given some well-defined model. These probabilities are based on long-run frequencies concerning what would happen over an infinite number of identical, repeated experiments. Within this framework, the statistical procedures in each of the three areas that we discussed are assessed in terms of their performance over the long run, and the associated probabilities pertain to the method rather than to the particular results that are obtained.

Savage (1962: 25-29) distinguished between the precision to be expected of an experiment before it is performed—*initial precision*—and the precision actually yielded when the experiment is performed—*final precision*. In these terms, classical procedures are evaluated entirely on the basis of initial precision. The "goodness" of a procedure is judged on the basis of what would happen over the long run. There is little basis for judging the "goodness" of a particular result obtained using that procedure, other than indirectly through its association with a good procedure. Thus while the classical procedures yield a summary of the data that were obtained, they do not enable us to say a great deal about those particular results. As Barnett (1982: 183) notes, the criticism that is directed at the classical approach in this regard is that "the *frequency* basis of a *procedure* has no relevance in assessing what we really know in some situation *after we have carried out that procedure.*" Yet it is the final precision that is the concern of the user of the findings. The user wants to know what to conclude or do on the basis of a particular finding; the classical approach stops short of providing much direction here.

FURTHER READING

The classical procedures discussed here can be found in most intermediate-level social statistics texts, such as Hays (1973). Criteria for choosing testing and estimation procedures are discussed in statistical theory texts such as Larson (1974), Lindgren (1976), and Silvey (1975). (The above texts all contain discussions of Bayesian procedures, as well.) The decision-theoretic basis of hypothesis testing is clearly expressed in Neyman (1950, 1952). Fisher's ideas of significance testing and his views of the Neyman-Pearson theory of hypothesis testing are outlined in Fisher (1973). Tukey (1960) discusses the issue of drawing conclusions versus making decisions in testing. Cox and Hinkley (1974) provide a detailed discussion of both types of tests. Seidenfeld (1979) contrasts the approaches of Neyman-Pearson and Fisher. Also, see Good (1983).

Recent debate and discussion of what significance and hypothesis tests do, and should do, can be found in the set of papers edited by Kadane (1976) and in the papers in a special issue of *Foundations of Probability and Statistics* (1977); both of these sets contain papers by Neyman. Morrison and Henkel (1970) contains a number of papers on the debate and controversy surrounding the use of testing in the social sciences. Overviews of the logic of the classical approach can be found in Barnett (1982) and Kyburg (1974: ch. 2). Additional references that compare features of the classical and Bayesian approaches are provided in Chapter 5 of this book.

NOTES

1. When the null hypothesis is a composite hypothesis, different sampling distributions could be constructed for each point in H_0. However, if the data required rejection of the simple null hypothesis that $\mu = \mu_0$, they will also require rejection of any simple null hypothesis in H_0. If the data do not require rejection of the null hypothesis that $\mu = \mu_0$, this is consistent with $\mu \leq \mu_0$.

2. If the investigator is concerned about making a Type II error only if Δ is substantially larger than Δ_0, the probability of error may be small for the values of Δ in which he or she is interested. It should be recognized, however, that rejection of H_0 in this test means only that one rejects the hypothesis that the true value of Δ is less than or equal to Δ_0. If the investigator is really interested in whether or not the true difference is less than or equal to some value that is greater than Δ_0, then rejection of the null hypothesis may not be a very informative outome.

3

Probability and
Its Interpretation

This chapter will be concerned with some basic ideas in probability theory. Specifically, the axiomatic basis of probability theory and certain definitions will be reviewed, and different meanings given to the notion of probability will be discussed. Consideration of this material is important because it is at this fundamental level that differences between the Bayesian and classical approaches arise. Whereas these two approaches accept the same axioms of probability, the underlying interpretations of probability are very different. As a consequence, the goals of statistical inference and the meaning of the results in classical and Bayesian statistics are different. Failure to recognize the differences at this level leads to confusion about the nature of Bayesian statistics.

AXIOMS AND DEFINITIONS

Most presentations of modern probability theory involve three basic axioms. These axioms impose certain restrictions on the numbers that we call probabilities. Before presenting these axioms, however, the idea of an *event* needs to be spelled out. Basically, an event is an outcome, or set of outcomes, of an experiment. Although the term "experiment" calls to mind a controlled laboratory or field experiment, the term is used here in a very general sense to denote a specified procedure that results in a single, well-defined outcome. A toss of a die, for example, is an experiment that results in one of six possible outcomes corresponding to the number of dots showing on the upper face when it comes to rest. If we are simply interested in the number of dots, and not in the particular position of the die, each of these six outcomes is an *elementary event*. Together they make the *event space* or *sample space* of the experiment.

These elementary events may be more elementary than is necessary for some purposes. One might only be interested in whether the die turns up with three dots or less. This is a *compound event* consisting of the individual elementary events of the die showing one, two, or three dots. If an event contains all possible elementary events, it is called the *universal event*. Two events are said to be *disjoint* if they do not share any of the elementary events; at most, only one of these disjoint events can occur in a single experiment.

The probability of an event E, denoted P(E), is a number that satisfies the following axioms:

(1) $P(E) \geq 0$, that is, the probability of an event is nonnegative.
(2) If U is the universal event, $P(U) = 1$.
(3) If events E_1 and E_2 are disjoint events, the probability of E_1 or E_2 occurring is the sum of $P(E_1)$ and $P(E_2)$.

These axioms are known as the "Kolmogorov axiom system"; there are some slightly different versions of the axioms in which certain theorems and axioms are interchanged; however, these differences are not important for this discussion. The important point to note here is that the axioms simply state certain properties that numbers called "probabilities" must have. They do not specify any particular interpretation of these numbers, nor any particular method of assigning probabilities.

The axiom system provides the basis for the calculus of probability. However, in order to proceed very far, certain definitions are required.

Two related definitions important for this discussion are those concerning independent events and conditional probability. Let the occurrence of both E_1 and E_2 be represented by E_1, E_2. Then E_1 and E_2 are defined to be independent if and only if

$$P(E_1, E_2) = P(E_1)P(E_2) \qquad [3.1]$$

That is, they are independent if and only if their joint probability is equal to the product of their individual probabilities or marginal probabilities. The conditional probability of E_1, given the occurrence of E_2 is written $P(E_1|E_2)$ and is defined

$$P(E_1|E_2) = \frac{P(E_1, E_2)}{P(E_2)} \qquad [3.2]$$

where $P(E_2) \neq 0$. It can be seen by substituting the right-hand side of equation 3.1 for $P(E_1, E_2)$ that the conditional probability $P(E_1|E_2)$ will be equal to the unconditional marginal probability $P(E_1)$ if and only if the two events E_1 and E_2 are independent.

At this point Bayes' theorem can be obtained in a few simple steps. It follows directly from the definition of conditional probability in expression 3.2. If we multiply both sides of that equation by $P(E_2)$ and reverse the right- and left-hand sides of the equality we get

$$P(E_1, E_2) = P(E_1|E_2)P(E_2)$$

Likewise, from the probability of E_2 conditional upon E_1, we get

$$P(E_2, E_1) = P(E_2|E_1)P(E_1)$$

The left-hand sides of these two equations are the same, being simply different notation referring to the same joint event. It follows then that

$$P(E_1|E_2)P(E_2) = P(E_2|E_1)P(E_1)$$

and

$$P(E_1|E_2) = \frac{P(E_2|E_1)P(E_1)}{P(E_2)}$$

This is a basic and very simple version of the theorem. The value of the theorem is that it provides the basis for revising $P(E_1)$ on the basis of new knowledge about the occurrence of E_2; the revised probability is $P(E_1|E_2)$. Discussion of the theorem and its use will be reserved for subsequent chapters where it will be considered in detail. The point of introducing it here is to show that it follows from some very basic ideas in probability theory. We turn now to the different meanings attached to probability.

INTERPRETATIONS OF PROBABILITY

As was mentioned in the previous discussion, the axioms of probability are simply mathematical requirements that numbers must satisfy in order to be called probabilities; the interpretation of such numbers is unspecified. We will briefly review an early interpretation that set the stage for much of the debate over probability, and three current interpretations. Although discussion of the interpretation of probability seems somewhat removed from the interests of evaluators and applied researchers, it is necessary for understanding the different aims and methods of classical and Bayesian statistics.

THE CLASSICAL SCHOOL

Consideration of the classical interpretation of probability is useful because many later developments in the different interpretations of probability occurred in response to classical ideas. The name "classical" is somewhat unfortunate here because it is not related to what we have referred to as the classical theory of statistics. Indeed, classical statistics is based on a different notion of probability, as will be seen.

This was the school of Bayes and Laplace, among others. Laplace (1951: 6) defined the probability of an event as the proportion of outcomes making up that event, to the total number of outcomes. Thus the probability of drawing a heart from a deck of cards is 13 divided by 52, or .25. This rests upon the assumption that the outcomes are all equally likely. The problem here is the following: How do we know that the outcomes are equally likely to begin with? One response to this was simply to point out the natural symmetry of the experiment and argue that the outcome must therefore be equally probable. This argument was criticized for its circularity (Barnett, 1982: 74). Another response to this was to rely on the principle of insufficient reason—if there is no reason to believe that one outcome is any more likely than any other, then we should assume that they are equally likely. Critics argued that this interpretation of probability was too subjective because it was based on what an individual believed or knew about the outcomes. Different individuals could disagree about the probability of the same event. In addition, inconsistencies could appear with different definitions of the outcomes. For example, if an outcome could be subdivided, then the probabilities would be different from those obtained without subdividing that outcome. In any case, probability interpreted in this way has a rather limited range of applications. It is relevant mainly to phenomena such as coin tosses, drawings from decks of cards, and so on, where some natural symmetry makes judgments of "equally likely" seem reasonable.

THE FREQUENCY SCHOOL

The development of the frequency school was, in part, a reaction to the subjectivity of the frequency interpretation. The focus of the classical school was on providing an objective, empirical basis for probability. Probability was viewed in terms of relative frequency. In other words, if an experiment is repeated n times, and event E occurs n_E times, then the probability of E is defined as $P(E) = n_E/n$. To be more precise, it is defined as the limiting value of this relative frequency as n approaches infinity. This idea was put forward in the mid-nineteenth century in criticism of the classical school and was given a systematic, mathematical treatment in the 1920s. The importance of this school is that it forms the basis for the classical theory of statistics. The reader will recall from the discussion in the previous chapter that in classical statistics, probability statements refer to what happens in the long run over identical repeated experiments.

There are two points to note here. First, probability is viewed as a feature of physical reality—it does not depend upon who is making the probability assessment. It is argued that we should all agree on the probability of an event defined in this way. Frequency probability is sometimes called objective probability for this reason. Second, probability can only be assigned to results of experiments that can, at least in principle, be conducted repeatedly. The holder of frequency views has little to say about the probability of unique nonrepeatable events. The frequency interpretation does not allow a researcher to assign a probability to treatment A outperforming treatment B in terms of some true difference $\Delta = \mu_A - \mu_B$. The idea of an infinite series of repeated experiments involving the two treatments in which μ_A exceeds μ_B in some proportion of these experiments seems a little farfetched. We might intuitively think in terms of how likely it is that treatment A is more effective than treatment B, but this notion of probability has no meaning in frequency terms.

The frequency school has been criticized, however, for being neither as empirical nor as objective as it is purported to be. To begin with, probability is defined in terms of a limit of an infinite series, yet as critics have noted, it is impossible to observe an infinite series in reality and to provide empirical confirmation. Winkler (1972: 14) describes the relative frequency definition as a conceputal rather than operational definition because it is not a realistic means for actually determining probabilities. Furthermore, it is not clear that subjectivity has been eliminated because the class of relevant experiments that determines n in the definition $P(E) = n_E/n$ is not always obvious. Leamer (1978) discusses the seemingly simple case of flipping a coin. Are all flips of the coin really identical repetitions of the same experiment? In one sense, if everything really were identical, the same outcome would occur. Indeed, many things are different on each flip, and in order to calculate a relative frequency we need to define the class of events over which the frequency is counted. The point is that "we may all agree on the class of events and in that sense have an 'objective' frequency, but that objectivity is something in us and not in nature" (Leamer, 1978: 25).

THE LOGICAL SCHOOL

A response to the frequency interpretation is to deny that probability is an empirically based statement about nature. This is expressed in the logical interpretation of probability often traced to the English economist John Maynard Keynes. Here probability is taken as representing a logical relation between a proposition and some evidence or body of knowledge. The idea is that we speak of propositions as probable in relation to our knowledge about them; probability is conditional upon our knowledge. Probabililty represents degree of belief; and for a given body of evidence, there is only one degree of probability that a given proposition may have conditional upon the evidence. Probability expresses a logical rather than an empirical relationship. For this reason, logical probability is sometimes also called "necessary probability."

This interpretation has been of somewhat more interest to philosophers than to statisticians. Yet, Jeffreys's influential work on Bayesian statistics is classified in this school, and the ideas of the logical school on representing a baseline or starting point in the accumulation of knowledge appear in many discussions of the application of Bayesian statistics.

THE SUBJECTIVE SCHOOL

Followers of the subjective school also deny that probability is an empirical statement about the world. De Finetti, a leading proponent of the subjective interpretation, writes in the preface of his two-volume work on probability (1974a: x):

> My thesis, paradoxically, and a little provocatively, but nonetheless genuinely, is simply this:
>
> PROBABILITY DOES NOT EXIST.
>
> The abandonment of superstitious beliefs about the existence of Phlogiston, the Cosmic Ether, Absolute Space and Time, . . . , or Fairies and Witches, was an essential step along the road to scientific thinking. Probability, too, if regarded as something endowed with some kind of objective existence, is no less a misleading misconception, an illusory attempt to exteriorize or materialize our true probabilistic beliefs.

As in the logical view, probability is seen as a degree of belief. However, in contrast to the logical view, such probability is not unique; it can vary from individual to individual. It is simply an expression of an individual's uncertainty about a relation or occurrence of an event. As Kingman (1975: 98) writes, "for De Finetti probability is not a scientific concept used to explicate observed physical phenomena, but a way in which individuals can communicate their assessments of uncertainty. It is essentially subjective, and any pretence at objectivity is misleading and dangerous." Although these ideas had been expressed in the early eighteenth century, and had been given formal treatment by De Finetti and Ramsey in the early part of this century, the subjective interpretation did not receive a great deal of attention until the publication of Savage's book in 1954. The subjective interpretation is the predominant one among Bayesians today, and it is the interpretation adopted in this book.

Here probability is conditional upon the individual. Edwards et al. (1963: 197) state that "any probability should, in principle, be indexed with the name of the person, or people, whose opinion it describes." In many instances people will agree on the probability of an event and, as we will see in subsequent chapters, the accumulation of data can serve to bring individuals with different initial beliefs into agreement. The main point here is that probability is seen not as an objective fact of nature, but rather as a subjective judgment that the individual makes, conditional on the individual's information and experience. Subjective probability is also referred to as personal probability.

It can be seen that the subjective interpretation of probability is broader than the frequency interpretation because it applies to non-repetitive as well as repetitive events. Because probability is a degree of belief, it can be applied to unique events. In addition, subjective probabilities can also be assigned to repetitive events; when available, relative frequencies may be an important part of the information that determines an individual's subjective probability. In some instances, frequency probabilities and subjective probabilities will be numerically the same. As will be seen, the subjective interpretation is closely tied to decision theory in that subjective probability is generally operationally defined in terms of an individual's choices in the face of uncertainty. This will be of interest to us because it links probabilistic information with its use in decision making.

Critics of the subjective view have raised a number of questions. To begin, why should subjective degrees of belief satisfy the axioms of probability? Furthermore, if they do satisfy the axioms, how can they be measured? Finally, even if they satisfy the axioms and can be measured, what relevance do subjective degrees of belief have for scientific investigation? We will look at the first question in this chapter, and address the remaining ones in subsequent chapters.

The main point here is that probability can be interpreted in a variety of ways. As one can imagine, these differences in interpretation have significant implications for statistical inference. Those holding frequency views will be concerned with phenomena that can be treated in long-run frequency terms, and their conclusions will be expressed in terms of what would happen over repeated experiments. Those holding logical or subjective views, on the other hand, will be concerned with degrees of belief and their relation to new evidence. The ramifications of these differences will be considered in some detail in the following chapters. Most Bayesians are adherents of the logical or subjective interpretations, whereas users of classical methods follow the frequency interpretation. As we shall see, persons holding frequency views can make only limited use of Bayes' theorem in statistical inference.

DEGREE OF BELIEF AND PROBABILITY

AXIOMS OF COHERENCE

An obvious question concerning the subjective interpretation is as follows: What justification is there for treating degrees of belief as probabilities? Although we often intuitively think in terms of how likely some event is and recognize different degrees of certainty, it is not immediately obvious that these beliefs should satisfy the axioms of probability. For a long time, no clear justification existed, and this was considered a weakness in the Bayesian argument, which involved subjective probabilities. Modern work, however, has established the existence of subjective probabilities on the basis of certain axioms of coherence pertaining to decision making. The problem is approached by considering what basic principles or rules a person should adhere to in making a choice, and then examining the implications of these principles. Formally, these principles are expressed as axioms that are used to derive certain theorems regarding judgments, preferences, and choices.

The problem is structured in the following way: We start at the point where an individual decision maker has identified alternative courses of action $a_1, a_2, \ldots, a_n$ among which he or she must choose. We also begin with a set of events $E_1, E_2, \ldots, E_n$ that could occur or become known subsequent to the acts. These events are sometimes referred to as "states of the world." A combination of an act a_i and an event E_j is a consequence c_{ij} of the decision. In nontrivial problems, the decision maker's preference for the different consequences will vary, and this should have some bearing on the act that is chosen. We will be interested in the type of problem where the occurrence of the events (and the consequences) is not known with certainty; this uncertainty should also have some bearing on the choice of an act. Starting at this point does not mean that the identification of acts and possible subsequent events is a simple or trivial matter. Much of what constitutes being a good decision maker lies in formulating the problem. In this chapter, however, we focus on the treatment of the decision subsequent to its formulation in the terms described.

The axioms of coherence, and theorems derived from them, concern how the individual's judgments of uncertainty for events, preferences for consequences, and choice of action fit together in a consistent fashion. There are various systems of axioms. Fishburn (1981), for example, provides a review of close to thirty different axiom systems; however, they all lead to the same general results. Basically, the axioms are statements of what most of us would regard as reasonable principles or requirements concerning preference, judgment, and action.

Edwards et al. (1965) informally discuss a number of axioms that appear in many of the axiom systems. One concerns *decidability*. Given any two consequences a and b, the decision maker should be able to tell whether he or she prefers a to b, prefers b to a, or is indifferent. Similar considerations apply to acts and judgments concerning the occurrence of events. Another axiom pertains to *transitivity*: If consequence a is preferred to consequence b, and consequence b is preferred to consequence c, then consequence a should be preferred to consequence c. Why is transitivity desirable? Suppose that the decision maker's preferences are intransitive and that instead of preferring a to c, he or she prefers c to a, and that he or she now holds c. Because b is more preferred, the decision maker should be willing to exchange c, plus some small amount of money at least, to obtain b. Likewise, he or she should be willing to exchange b plus some amount of money to obtain a. Then, if the decision maker's preferences are intransitive and he or she prefers c to a, he or she should be willing to exchange a plus some amount of money for c. At this point, the decision maker is back where he or she started from *minus* the money paid to make the exchanges. The cycle could now be repeated over and over until the individual eventually goes bankrupt or revises his or her preferences. Because of his or her intransitivity the individual can be used as a "money pump." Most of us would see this as undesirable.

Yet another axiom concerns *dominance*. If no matter what event occurs, the consequences of act a_1 are at least as desirable as the consequences of act a_2, and with the occurrence of at least one of those events, the consequence obtained with a_1 is more desirable than that obtained with a_2, then the decision maker should prefer a_1, and a_2 and other dominated acts could be eliminated from consideration. A final axiom discussed by these authors involves the *sure thing principle*, which says that when choosing between two possible actions, consequences that do not vary with your choice should not affect your choice. In other words, if acts a_1 and a_2 lead to consequences that are equally preferable under some event or state E_i, then these consequences should not affect your choice.

Certain axioms typically involve qualitative assessments on the part of the decision maker regarding preferences for consequences and judgments concerning the occurrence of uncertain events. LaValle (1978) describes these assessments as being of the "oculist variety"; they are of a "preferred to-not preferred to" or "greater than-not greater than" nature, and in the same way that the oculist's comparisons lead to a quantitative index of visual acuity, these qualitative assessments of preferences and judgments lead to quantitative scalings. The important point is that, as Lindley (1976: 364) writes, the axioms "express essentially *simple* requirements on the behavior of the decision-maker, in the sense that any decision-maker who violated them would feel that he had made such an elementary error that he could hardly be forgiven." Although we may not always behave as these axioms prescribe, once the violation is pointed out, most of us would feel compelled to modify our behavior.

Given these axioms, it can be proved that the decision maker's uncertainties concerning the occurrence of the events or states are described by probabilities satisfying the axioms of probability theory; the conditional probability rule, and hence Bayes' theorem, can also be derived. Furthermore, it can be proved that the decision maker's preferences for the consequences are described by a quantitative utility function, and that the decision maker should choose among acts in such a way as to maximize expected utility, where the expected utility of an act is the weighted average of the utility of the possible consequences of an act with weights being the subjective probabilities associated with the states. The expected utility EU of an act a_i can be written

$$EU(a_i) = \sum_{j=i}^{n} P(E_j)U(c_{ij}) \qquad j = 1, 2, \ldots, n$$

where $P(E_j)$ represents the individual's subjective probability for event E_j, and $U(c_{ij})$ represents his or her utility for consequence c_{ij}. The choice of the act that maximizes expected utility is not based on any long-run considerations; it simply follows from the axioms that it is the preferred act in that particular decision problem. The reason for the use of the term "expected" is mainly historical, and its use here does not refer to an expectation over repeated decisions. This is sometimes referred to as expected utility (EU) theory or, to emphasize the subjective probabilities involved, subjective expected utility (SEU) theory.

Lindley (1971: v) emphasizes the prescriptive nature of these theorems in his presentation of the theory of decision:

The main conclusion is that there is essentially only one way to reach a decision sensibly. First, the uncertainties present in the situation must be quantified in terms of values called probabilities. Secondly, the various consequences of the courses of action must be similarly described in terms of utilities. Thirdly, that decision must be taken which is expected—on the basis of the calculated probabilities—to give the greatest utility. The force of "must," used in three places there, is simply that any deviation from the precepts is liable to lead the decision-maker into procedures which are demonstrably absurd—or as we shall say, incoherent.

The concept of coherence is important in the subjective argument. In an early treatment, De Finetti (1964: 151) writes that the condition of coherence "obliges us to take care in evaluating probabilities not to allow an adversary who bets against us the possibility of winning with certainty, whatever may be the event that occurs, by a judicious combination of stakes on the various events." It can be shown that coherence requires that one's subjective probabilities obey the probability axioms. More generally, the idea is that one's preferences, judgments, and actions should fit together or cohere (Lindley, 1971, 1972). Coherence is the primary standard of adequacy within the subjectivist framework. Regarding subjective probabilities, the individual is free to assign probabilities to events in whatever way seems reasonable provided that these assignments are coherent, which requires that they satisfy the probability axioms.

COMMENTS ON THE AXIOMATIC BASIS

Normative Versus Descriptive Arguments

The theory is sometimes challenged on the grounds that people frequently violate the axioms in actual decision making. It is important to recognize, however, that the argument here is normative rather than descriptive. It specifies the form that one's beliefs, values, and choices should take if one accepts the axioms of coherence as reasonable. Although one can easily find instances of incoherence in everyday behavior, this does not discredit the theory any more than the fact that people make computational errors in arithmetic somehow discredits the rules for arithmetic operations. From the normative point of view, incoherence does not castigate the theory; rather, it embarrasses the individual (Barnett, 1982: 110). Indeed, the frequent occurrence of incoherence argues for the need for a formal theory to provide guidance in making important decisions. As Smith (1984: 248) writes, "The more prone people are to make mistakes the more need there is for the formal prescription, both as a tool for discovering the kinds of mistakes and distortions to which we are prone in *ad hoc* reasoning and for improving strategies for structuring our thoughts about problems." A much-discussed criticism of the theory on the basis of observed incoherence can be found in Allais (1953), and a number of related references are provided at the end of this chapter.

A related, but somewhat different challenge to the theory has its basis in psychological research on decision making. The SEU theory has been a standard point of reference or model in psychological theories of decision making. The theory does have some general appeal as a descriptive theory. It embodies the rather commonsense notion that people act in ways that they believe are likely to lead to results that they consider to be desirable, and SEU-like ideas appear implicitly or explicitly throughout a wide range of psychological, sociological, economic, and philosophical theories of human behavior, as well as in everyday motivational attributions and explanations of behavior. Yet, recent research in this area has been critical of the SEU theory as a descriptive model of human performance. Following the 1974 paper by Tversky and Kahneman, in which they identify certain cognitive simplifying operations or heuristics that lead to biases in judgments, a large number of studies have focused on demonstrating suboptimal performance with respect to normative theories, and summaries of this literature lead to fairly pessimistic conclusions about human performance (Slovic et al., 1977; Nisbett and Ross, 1980). Much of this pertains to SEU theory as a descriptive theory, but it has implications for the normative aspect of the theory as well because the argument can be made that humans have such limited cognitive abilities and such biased methods of processing information that they can neither provide meaningful input for the normative framework nor carry out the operations required.

However, there continues to be a debate in this area between what Jungerman (1983: 79) calls the "pessimists" and the "optimists" concerning human performance: "While one side provides evidence that people do not behave as the model predicts, the other side demonstrates that with different theoretical assumptions or experimental settings the same behavior might not violate the model at all." Tversky and Kahneman (1981: 458), for example, present research findings showing that subjects' preferences change depending upon how the decision problem is framed, and argue that the findings "raise doubt about the feasibility and adequacy of the coherence criterion." Yet Berkeley and Humphreys (1982) offer an alternative interpretation of subjects' performance that is consistent with SEU theory. They argue that conclusions about biases and errors in this area of research rest upon the assumption that the subject's internal representation of the decision problem is the same as that of the experimenter, and contend that this assumption is unjustified. They go beyond this to suggest that the willingness to accept conclusions about bias is itself a bias on the part of experimenters, which they label the "bias heuristic."

In addition, even if one does accept the experimenter's assumptions, other critics, such as Edwards (1983), argue that the findings pertain largely to judgment and decision tasks that are abstract, divorced from real-life settings, and performed without tools. In commenting on the nature of the tasks, Phillips (1983: 533) writes, "The heuristics literature abounds with abstract problems, hypothetical gambles, imaginary games with abstract probabilities of winning or losing amounts that materialise out of thin air, imagined balls and urns with samples generated by unseen forces, life and death decisions, choices between large amounts of money and stated probabilities that differ by only 0.01 and 0.05, unfamiliar contexts" and questions the generalizability of findings based on tasks such as these.

In a similar critical vein Christensen-Szalanski and Beach (1984) identify a "citation bias" in the literature in this area, with a study of citation frequencies revealing that studies showing poor performance are cited more frequently than studies showing good performance. In part, this may be due to the fact that deviations from good performance are considered more intriguing and remarkable than good performance itself. As Edwards (1983: 507) writes, "If someone says '2 + 2 = 4,' that isn't psychology; it is just arithmetic. But '2 + 2 = 5' is psychology. If enough experimental subjects say it often, it will be a finding, and the experimental and theoretical literature will burgeon." Whatever the reason, the emphasis in the literature can give the impression that human judgment and performance is plagued by biases and this can lead to inappropriate overgeneralization. Berkeley and Humphreys (1982) note that this overgeneralization has been especially true in writings outside of the psychological decision theory literature; there the "bias" conclusion has been discussed as established fact.

The point of discussing this material here is that it is far from clear that findings in this area concerning limitations in human performance preclude the prescriptive use of the theory. There is no doubt that human judgment and performance may be less than perfect in many situations; however, as Beach et al. (1984: 1) comment, "Imperfection does not necessarily imply shambles." Furthermore, research on performance that does focus on individuals dealing with real-life tasks with the proper training, tools, and experience provides evidence of good performance; for example, see the review by Murphy and Winkler (1984).

Rational Man

It is also important to recognize that the theory does not rest upon the assumption of an idealized, omniscient, "rational man." It merely prescribes how an individual should make choices in a way that will be logically consistent with what the individual believes and values at the time of making the decision. As we have pointed out, the theory does not presuppose that an individual behaves according to the axioms. The purpose of the theory and the associated technology is to lay out reasonable principles and aid the person who accepts these principles in performing in a manner that will be consistent with them.

Inference and Decision Making

A final comment is that while we have referred to an axiomatic approach that provides for the simultaneous justification of subjective probability and utility, there are alternative axiom systems where these are derived separately. The advantage of the simultaneous treatment is that the relationship of beliefs, values, and actions can be seen, and this facilitates consideration of decision making in our subsequent discussion. The argument should not be interpreted to mean that subjective notions of probability apply only when one has a well-specified decision problem. Indeed, much of the treatment of subjective probabilities in Bayesian statistics concerns inference, not decision making. However, if these inferences are to be used to guide action at some point, then the decision-making framework can be useful.

FURTHER READING

Discussion of the Kolmogorov axioms can be found in most statistics texts. Some Bayesians prefer to work with an alternative axiom system described by Renyi (1970). In this system probabilities are explicitly treated as conditional in the basic axioms and, while not inconsistent with the Kolmogorov axioms, they are better suited for expressing Bayesian ideas about changes in belief with changes in conditioning events. Lindley (1965a) uses the Renyi axioms, and Novick and Jackson (1974) discuss the two axiom systems. Good (1982) contains a brief discussion of different axiom systems.

Comparisons of the different interpretations of probability discussed in this chapter can be found in Barnett (1982), De Finetti (1978), Good (1950, 1959, 1965, 1983), Hartigan (1983), Kyburg and Smokler (1980a),

Lee (1971), and Savage (1972). Fine (1973) provides an extensive review of a number of theories of probability, including theories not discussed here. References to primary sources can be found in these reviews. Kyburg and Smokler (1964, 1980a) provide a collection of papers that have been influential in the development of the subjective interpretation of probability.

There are a number of normative subjective expected utility theories with different axioms. Fishburn (1981) reviews these theories and recommends the theories of Savage (1972) and Pratt et al. (1964, 1965) on the basis of a number of criteria. Pratt et al. provide a useful informal treatment as well as a formal argument. Lindley (1971a) reviews a number of theories, and Lindley (1971b) provides an introductory presentation in which he derives the Renyi probability axioms. See also discussions of the axioms in Berger (1980), DeGroot (1970), Eells (1982), Fine (1973), Fishburn (1970), Krantz et al. (1971), LaValle (1970, 1978), Luce and Suppes (1965), MacCrimmon and Larsson (1979), and Winkler (1972). Lindley (1982) provides a brief overview of the coherence argument.

Bayesian responses to the argument in Allais (1953) can be found in Savage (1972) and Raiffa (1968); additional papers concerning the Allais argument are collected in Allais and Hagen (1979). Slovic and Tversky (1974) present an Allais-type argument. See also the Ellsberg (1961) and Fellner (1961) criticisms and the responses by Raiffa (1961) and Roberts (1963).

Readers interested in further discussion of recent decision research might wish to review, in addition to those references already discussed, the review papers by Einhorn and Hogarth (1981), Pitz and Sachs (1984), and Wallsten and Budescu (1983), the monographs by Hogarth (1980) and Wright (1984), and edited collections by Beach et al. (1980), Feather (1982), Humphreys et al. (1983), Kahneman et al. (1982), Scholz (1983), and Sjöberg et al. (1983). The implications of decision research for descriptive and normative theory are also considered in Cohen (1981) and Kyburg (1983) with comments by numerous discussants.

4

Bayes' Theorem

In this chapter, we will discuss Bayes' theorem in some detail, and illustrate how it is used in revising probabilities. The role of the component terms will be examined, and different versions of the theorem for events and for random variables will be presented. Finally, we show how it can be applied to revision of belief about population or process parameters, and this provides the framework for subsequent discussion of specific Bayesian techniques. As was pointed out in the previous chapter, Bayes' theorem follows directly from the axioms of probability and the definition of conditional probability. The mathematical basis of the theorem is not controversial. In discussing the theorem here, we will downplay the controversy concerning its application by dealing with examples where frequency data in the form of rates and proportions are available. In this situation, advocates of the different interpretations of probability might all arrive at numerically equivalent probabilities. Our goal is to describe the workings of the theorem and introduce some notation to be used in the following chapters. Aspects of the controversy surrounding application will be discussed at the end of the chapter.

BAYES' THEOREM FOR EVENTS

In the last chapter we presented a version of Bayes' theorem for events E_1 and E_2:

$$P(E_1 | E_2) = \frac{P(E_2 | E_1)P(E_1)}{P(E_2)} \qquad [4.1]$$

where $P(E_2) \neq 0$. This allows us to revise the probability $P(E_1)$ to $P(E_1 | E_2)$ given the new data or knowledge that E_2 has occurred. Knowing that E_2 has occurred, we can determine the revised probability of E_1 if we have values for the terms on the right-hand side of equation 4.1. In a narrow sense Bayes' theorem is simply a rule for calculating conditional probabilities; however, in a broad sense it is the essence of a theory of learning from new data.

In talking about events in the abstract, however, it is difficult to see much practical value in the theorem. It is not clear why one would happen to know all the terms on the right-hand side of the equation, but not know the term on the left-hand side. A simple example can help clarify this. Consider the problem of determining the probability that an individual has a disease, given the result of a diagnostic test administered in a screening program. Let D represent the event "the individual has the disease," and let $\overline{D}$ represent the event "the individual does not have the disease." Also, let T^+ represent the event "positive test result" and T^- represent the event "negative test result." Suppose that we wish to determine the probability of the event "individual has the disease," given the event "positive test result." The probability of interest here is the conditional probability $P(D | T^+)$. This can be expressed in terms of Bayes' theorem as follows:

$$P(D | T^+) = \frac{P(T^+ | D)P(D)}{P(T^+)} \qquad [4.2]$$

If we know the values of the terms on the right-hand side of the equation, then the desired probability can be obtained. The first term in the numerator $P(T^+ | D)$ is the probability of obtaining a positive result if the individual does indeed have the disease. Persons familiar with diagnostic testing will recognize this as an expression of the true positive rate. The true positive rate is usually determined from previous studies of the test's validity and diagnostic value. The second term in the numerator $P(D)$ is the unconditional probability that the person has the disease. This is often determined from previous research on the base rate or frequency of occurrence of the disease in the population, and is an expression of the prevalence of the disease. The denominator $P(T^+)$ is the probability of obtaining a positive test result. It is an expression of the rate at which this result occurs in the population. Although it is conceivable that this rate could be available from previous research, it is usually not available directly. However, upon closer examination it can be seen to be made up of certain components that may be more readily available.

To show how $P(T^+)$ may be reexpressed, we begin by noting that the events D and $\overline{D}$ are disjoint or exclusive and they are exhaustive—all persons either have the disease or do not have it. Persons showing a positive test result can be conceptually separated into two groups: those who show a positive test result and have the disease, and those who show a positive test result and do not have the disease. Thus the event T^+ can be seen to occur when either of the joint events T^+, D and $T^+, \overline{D}$ occur. The events T^+, D and $T^+, \overline{D}$ are exclusive, and by Axiom 3, discussed in the previous chapter,

$$P(T^+) = P[(T^+, D^+) \text{ or } (T^+, \overline{D})] = P(T^+, D) + P(T^+, \overline{D})$$

Using the conditional probability rule we can write

$$P(T^+, D) = P(T^+ | D)P(D)$$

and

$$P(T^+, \overline{D}) = P(T^+ | \overline{D})P(\overline{D})$$

Thus we have

$$P(T^+) = P(T^+ | D)P(D) + P(T^+ | \overline{D})P(\overline{D})$$

Entering this into equation 4.2 we obtain the following statement of Bayes' theorem:

$$P(D | T^+) = \frac{P(T^+ | D)P(D)}{P(T^+ | D)P(D) + P(T^+ | \overline{D})P(\overline{D})} \qquad [4.3]$$

Instead of $P(T^+)$ in the denominator, we now have a number of terms there. How does this help us? Note that the first set of terms in the denominator is the same as the set of terms in the numerator, which we have already discussed. The second set is made up of $P(T^+|\overline{D})$, the probability of showing a positive result when the disease is not present (the false positive rate), which again is obtained from validity studies of the test; and $P(\overline{D})$, the probability of not having the disease, which is simply $1 - P(D)$.

Consider how this works with some numbers. Suppose that we have a test with the following characteristics for the population to which the individual under consideration belongs:

Probability of True Positive Result $P(T^+|D) = .75$
Probability of False Positive Result $P(T^+|\overline{D}) = .20$

Furthermore, suppose that from the prevalence rate, $P(D) = .07$ and $P(\overline{D}) = 1 - P(D) = .93$. Then we have

$$P(D|T^+) = \frac{(.75)(.07)}{(.75)(.07) + (.20)(.93)} = .22$$

On the basis of the test result, the probability that the individual has the disease should be revised from .07 to .22.

The probability that the individual does not have the disease can simply be obtained by subtraction: it is $1.00 - .22 = .78$. This could also be calculated using Bayes' theorem. In this case

$$P(\overline{D}|T^+) = \frac{P(T^+|\overline{D})P(\overline{D})}{P(T^+|D)P(D) + P(T^+|\overline{D})P(\overline{D})} \qquad [4.4]$$

and therefore

$$P(\overline{D}|T^+) = \frac{(.20)(.93)}{(.75)(.07) + (.20)(.93)} = .78$$

which agrees with the value obtained by subtraction. The main point here is that given that a positive test result was obtained, Bayes' theorem allows us to use knowledge of the test characteristics and the prevalence of the disease to determine the probability of interest.

In more general Bayesian terminology, $P(D)$ and $P(\overline{D})$ are called *prior probabilities*, or sometimes, simply *priors*. They are the probabilities that the individual does or does not have the disease, prior to obtaining the test result or new data. $P(T^+|D)$ and $P(T^+|\overline{D})$ are called *likelihoods*. They are the probabilities of obtaining the observed data (in this case, the positive test result) given the unknown state or condition of the individual (in this case, diseased or nondiseased). It is through the likelihoods, and only through the likelihoods, that data enter the Bayesian revision process. You will note that the likelihoods only involve the datum that was actually observed—neither equation 4.3 nor 4.4 involves T⁻, the negative test result. Of course, if a negative test result has been obtained, Bayes' theorem could be used to revise the prior probabilities in light of this; the appropriate likelihoods in this case, however, would be $P(T^-|D)$ and $P(T^-|\overline{D})$ rather than those used in equations 4.3 and 4.4. $P(D|T^+)$ and $P(\overline{D}|T^+)$ are called *posterior probabilities*, or *posteriors*. They are the probabilities of the unknown states of interest *posterior* to obtaining the data or datum. The prior probability and the likelihood in the numerator of Bayes' theorem are the terms that basically "do the work" in the revision process. The denominator remains constant for $P(D|T^+)$ and $P(\overline{D}|T^+)$, as can be seen in equations 4.3 and 4.4; it simply ensures that the posterior probabilities for the exhaustive set of events sum to one as required by Axiom 2. In some instances, the numerator has some value beyond simply standardizing the posterior probabilities. In this problem, the numerator expresses $P(T^+)$, the probability of obtaining a positive test result. Given the emphasis on diagnosis here, this may not be of interest. In other types of problems, however, prediction of future observations may be important.

The components of Bayes' theorem and the calculations for this example are displayed in Table 4.1. Starting on the left, we have the events and the priors. Next, we have the likelihoods for the particular datum that was observed. The priors and likelihoods are then multiplied together and divided by the sum of these products to yield the posteriors. Because the denominator in Bayes' theorem is the sum of the priors times the likelihoods for each of the events, this standardizes the posteriors so that they total to one. Thus, given the priors and the likelihoods, the posteriors or revised probabilities are easily obtained through the use of Bayes' theorem.

BAYES' THEOREM IN
ODDS-LIKELIHOOD RATIO FORM

When there are two events of interest such as disease and no disease, a slightly different version of Bayes' theorem is sometimes used. If we

TABLE 4.1
Components of Bayes' Theorem for Diagnostic Problem

Event	Prior Probability	Likelihood	Prior x Likelihood	Posterior Probability
Individual has the disease (D)	$P(D) = .07$	$P(T^+|D) = .75$	$.07 \times .75 = .053$	$.053/.239 = .22$
Individual does not have the disease ($\overline{D}$)	$P(\overline{D}) = .93$	$P(T^+|\overline{D}) = .20$	$.93 \times .20 = .186$	$.186/.239 = .78$
	Sum = 1.00	No Constraint on Sum	Sum = $P(T^+)$ = .239	Sum = 1.00

divide equation 4.3 by equation 4.4 the denominator in the two equations cancels out and we can write the result in the form of ratios:

$$\frac{P(D|T^+)}{P(\overline{D}|T^+)} = \frac{P(T^+|D)}{P(T^+|\overline{D})} \times \frac{P(D)}{P(\overline{D})} \qquad [4.5]$$

This says that the ratio of the posterior probabilities is the product of the ratio of the prior probabilities and the ratio of the likelihoods.

The relative probabilities of two disjoint events are sometimes expressed in terms of odds. The odds Ω in favor of an event E_1 over another event E_2 are $\Omega = P(E_1)/P(E_2)$. Odds are most commonly calculated when the two events are exhaustive. Then $P(E_2) = 1 - P(E_1)$ and the odds in favor of E_1 are $\Omega = P(E_1)/[1 - P(E_1)]$. If $P(E_1) = .75$, and $P(E_2) = .25$, the odds are $.75/.25 = 3.0$. This is read as odds of "three to one" in favor of E_1, and is sometimes written 3:1. Given the odds Ω, the individual probabilities can be recovered as follows: $P(E_1) = \Omega/(\Omega + 1)$ and $P(E_2)$ can be obtained by subtraction. Returning to equation 4.5, $P(D)/P(\overline{D})$ represents the prior odds in favor of the person having the disease, and $P(D|T^+)/P(\overline{D}|T^+)$ represents the posterior odds. $P(T^+|D)/P(T^+|\overline{D})$ is the likelihood ratio. Using the values for the probabilities and likelihoods that we discussed earlier, we can calculate the posterior odds in favor of disease, given the positive test result, as follows:

$$\frac{P(D|T^+)}{P(D|T^-)} = \frac{.75}{.20} \times \frac{.07}{.93} = 3.75 \times .075 = .28$$

This checks with the ratio of our previously calculated posterior probabilities: $.22/.78 = .28$.

When we are mainly interested in the relative probability of two events, this form of Bayes' theorem saves us the step of dividing by the sum of the likelihoods to standardize the posteriors. This form of the theorem is often used in comparing the relative probability of two hypotheses in light of new data. If H_1 and H_2 represent disjoint simple hypotheses, and D represents data, we can write

$$\frac{P(H_1|D)}{P(H_2|D)} = \frac{P(D|H_1)}{P(D|H_2)} \times \frac{P(H_1)}{P(H_2)}$$

Thus the posterior odds favoring H_1 over H_2 is the product of the prior odds and the likelihood ratio. Bayes' theorem in this form can be used in Bayesian hypothesis testing, if desired. The likelihood ratio is called the *Bayes factor* in some discussions (Good, 1965: ch. 5). This approach can also be applied when one or both of the hypotheses are composite, although the form of the likelihood ratio is somewhat more complex; we will return to this in a following chapter. The odds-likelihood ratio form of the theorem is sometimes written as $\Omega'' = LR \times \Omega'$ where Ω' and Ω'' are the prior and posterior odds, respectively, and LR is the likelihood ratio.

THE ACCUMULATION OF INFORMATION

Bayes' theorem can be applied repeatedly to revise probabilities as more information is obtained. Suppose that the screening test is followed by an examination for the presence or absence of a particular symptom of diagnostic value. Let S^+ represent the presence of the symptom and S^- the absence of the symptom. From the previous section, the probability that the individual has the disease has been revised on the basis of the test result. Bayes' theorem can now be used again to revise these posterior probabilities on the basis of the examination results.

Consider the case where the symptom is observed. The problem is to determine the probability that the individual has the disease given that the individual has a positive test result and now shows the symptom. This probability is $P(D|T^+, S^+)$, and it can be calculated using Bayes' theorem:

$$P(D|T^+, S^+) = \frac{P(S^+|D, T^+)P(D|T^+)}{P(S^+|D, T^+)P(D|T^+) + P(S^+|\bar{D}, T^+)P(\bar{D}|T^+)} \quad [4.6]$$

The terms $P(D|T^+)$ and $P(\overline{D}|T^+)$ are simply the posterior probabilities obtained in equations 4.3 and 4.4. For this second revision, however, they are now the priors that will be revised on the basis of the new data. The forms prior and posterior are relative terms—they are relative to the data being considered. It is sometimes said that "today's posteriors are tomorrow's priors." Now the likelihoods of interest are as follows: the probability that the individual shows the symptom, given that he or she has the disease and has a positive test result, and the probability that the individual shows the symptom when he or she does not have the disease and has a positive test result.

Suppose that it has been established on the basis of previous research that $P(S^+|D, T^+) = .65$, and that $P(S^+|\overline{D}, T^+) = .15$. Then, with $P(D|T^+) = .22$ and $P(\overline{D}|T^+) = .78$, calculated previously, we enter these values in equation 4.6 and obtain

$$P(D|T^+, S^+) = \frac{(.65)\,(.22)}{(.65)\,(.22) + (.15)\,(.78)} = .55$$

and

$$P(D|T^+, S^+) = \frac{(.15)\,(.78)}{(.65)\,(.22) + (.15)\,(.78)} = .45$$

Thus on the basis of both the positive test result and the observation of the symptom, the probability that the individual has the disease has been revised upward from the initial value of .07 to .55.

One point to note in equation 4.6 is that we did not work the simple likelihoods $P(S^+|D)$ and $P(S^+|\overline{D})$; instead, we had to use likelihoods that were also conditioned on the positive test result, $P(S^+|D, T^+)$ and $P(S^+|\overline{D},T^+)$. This is because the probability of showing the symptom given the disease or absence of the disease may be different for individuals with positive test results than it is for the population as a whole. If the occurrence of the symptom and the positive test result were positively associated, then these likelihoods would be different. One can see that the likelihood term can become increasingly complex (in terms of the conditioning considerations) as information is accumulated. The situation is simplified considerably when the observations, or pieces of data, are conditionally independent. Events E_1 and E_2 are defined to be conditionally independent given E_3 if

$$P(E_1, E_2|E_3) = P(E_1|E_3)P(E_2|E_3)$$

This parallels the definition of (unconditional) independence given in Chapter 3. If, in our present example, S^+ and T^+ were conditionally independent given D and conditionally independent given $\overline{D}$, then it could be shown that

$$P(S^+|D, T^+) = P(S^+|D)$$

and

$$P(S^+|\overline{D}, T^+) = P(S^+|\overline{D})$$

which means that we could simplify the likelihoods in equation 4.6. In problems of statistical inference that will be discussed in the following chapters, the likelihoods derive from some model describing the probabilities of the observations as functions of the model parameters. The problems that will be considered all involve the assumption that the observations are generated independently, conditional upon the particular values of the parameters in that model. This helps simplify the analysis; however, simply assuming conditional independence does not make it true. As in any analysis, the appropriateness of the assumptions needs careful consideration.

Note that the use of Bayes' theorem is not limited to problems involving dichotomous events such as disease/no disease, positive test result/negative test result, symptom/no symptom. The subsequent discussion will be concerned with the revision of probabilities over many events.

BAYES' THEOREM FOR RANDOM VARIABLES

The discussion of Bayes' theorem to this point has concerned simple discrete events. Although the forms of the theorem discussed above help to give one a feel for how Bayes' theorem works, they cannot be applied very easily to many of the important problems of statistical inference. In order to address these problems we need to consider the use of Bayes' theorem for random variables. Both discrete and continuous random variables will be considered, as well as a form of the theorem that enables us to make inferences about unknown parameters on the basis of observed random variables.

THE DISCRETE CASE

A random variable is a function that associates a real number with each of the elementary events in the event space. We can write $X(E) = x$ where E is the event and x is the value associated with that event by the random variable X. Suppose that we perform a simple experiment in which we toss three fair coins—a penny, a nickle, and a dime—and record in that order whether heads or tails appear. The elementary events making up the event space, or sample space, are HHH, HHT, HTH, HTT, THH, THT, TTH, TTT where H represents the occurrence of heads and T the occurrence of tails. Given that the coins are fair and that the tosses are independent, the probability of each outcome is $.5 \times .5 \times .5 = .125$. Now suppose that we assign a number to each outcome indicating how many heads are observed. We thus have a function X, say, that assigns a number to each of the elementary events as follows: $X(HHH) = 3$, $X(HHT) = 2$, $X(HTH) = 2$, $X(HTT) = 1$, $X(THH) = 2$, $X(THT) = 1$, $X(TTH) = 1$, $X(TTT) = 0$. Here, X is a discrete random variable with only a finite number of values. Given the probabilities of the elementary events, we can determine the probability that X takes on any particular value x. We will write this $P(X = x)$. For example, X has the value 0 only when the event TTT occurs. $P(X = 0) = P(TTT)$ and is therefore .125. X takes on the value 1 when any of the events HTT, THT, or TTH occur. Because these events are disjoint or exclusive, $P(X = 1) = P(HTT) + P(THT) + P(TTH) = .125 + .125 + .125 = .375$. $P(X = x)$ could be determined for other compound events in a similar manner.

In the subsequent discussion we will want to refer to the distribution of probability over the possible values of X, rather than simply to the probability that X has some specific value x. To do this, I will use a function that assigns a probability to each x. This probability function for the discrete random variable X will be written $p(x)$. That is, $p(x) = P(X = x)$ for all values of x. This function might simply be a listing that matches each x with a probability. However, in the statistical problems that we will be considering, the function will be an algebraic rule or formula that expresses the probability as a function of x. In some instances, we will talk about $p(x)$ for some particular value $X = x$. In other instances, we will want to refer to the distribution of probability over the possible values of X rather than simply to the probability that X has some particular value x, and we will use $p(x)$ to refer to this entire distribution. We will often refer to $p(x)$ as a probability distribution or, simply, as a distribution.

Bayes' Theorem

Suppose that we have two random variables. From the definition of conditional probability we can write

$$P(Y = y \mid X = x) = \frac{P(Y = y, X = x)}{P(X = x)} \qquad [4.7]$$

where $P(Y = y \mid X = x)$ is the conditional probability that Y has the value y given that X has the value x, and where $P(Y = y, X = x)$ is the joint probability that Y has the value y and X has the value x. The conditional probability $P(X = x \mid Y = y)$ can be expressed in an analogous manner. From this it is a short step to Bayes' theorem for random variables:

$$P(Y = y \mid X = x) = \frac{P(X = x \mid Y = y)P(Y = y)}{P(X = x)} \qquad [4.8]$$

We could use this formula to calculate the probability of a particular value of Y, given some particular value of X.

Using the probability function notation, we can rewrite equation 4.7 as

$$p(y \mid x) = \frac{p(x, y)}{p(x)} \qquad [4.9]$$

where $p(y \mid x)$ is the conditional probability function of Y given a particular value of X, $p(x, y)$ is the joint probability function of X and Y, and where $p(x) \neq 0$. In addition, we can rewrite Bayes' theorem in equation 4.8 as

$$p(y \mid x) = \frac{p(x \mid y)p(y)}{p(x)} \qquad [4.10]$$

For a particular value of X, we can think of Bayes' theorem in 4.10 as a rule for obtaining the entire conditional distribution over the range of Y. If Y can take on n discrete values $y_1, y_2, \ldots, y_n$, then

$$p(x) = \sum_{i=1}^{n} p(x \mid y_i)p(y_i) \qquad [4.11]$$

Joint, Marginal, and Conditional
Probability Distributions

At this point it is useful to look at a simple example involving joint, marginal, and conditional probability distributions and their relationship through Bayes' theorem. A survey of directors of community mental health centers in a six-state area is conducted in which the directors are asked to rate the importance of a number of potential goals for their center. Each goal is rated on a three-point scale: 1—not important, 2—moderately important, 3—very important. The centers are also classified on various characteristics, one of which is whether the center is located in an urban or a rural setting.

Consider the ratings for a particular goal, cross-classified by setting. We base the probability that an individual randomly drawn from this group will make a particular rating, and will be from a particular setting, on the proportion of respondents in each rating/setting category. The setting is treated as a random variable X that takes on the value of 1 if the setting is urban, and 2 if it is rural. The rating of that particular goal is treated as a random variable Y that takes on values of 1, 2, or 3 as described above.

The joint and marginal distributions of X and Y are shown in Table 4.2. The probabilities in the cells are the probabilities of the joint occurrence of particular values of X and Y, and they make up the joint distribution $p(x, y)$. The probabilities on the right-hand margin make up the unconditional marginal distribution $p(y)$, and the probabilities along the bottom make up the unconditional marginal distribution $p(x)$. The name "marginal" arises from the fact that these probabilities appear on the margins of this table of joint probabilities.

The conditional distribution $p(y|x)$ when $X = 1$, for example, is the distribution of probability over the different response categories, given that the center director is from an urban center. Using equation 4.9 the values are calculated to be

$$P(Y = 1|X = 1) = .163$$
$$P(Y = 2|X = 1) = .279$$
$$P(Y = 3|X = 1) = .558$$

Basically, in obtaining the conditional distribution $p(y|x)$ for $X = 1$, we are lifting out a column of the table and rescaling the probabilities so that they sum to one. The conditional distribution $p(x|y)$ for some particular value of Y could be found in a similar manner; here we would

TABLE 4.2
Joint and Marginal Probabilities of
Setting (X) and Rating (Y)

		Setting (X) 1	2	
	1	.07	.12	.19
Rating (Y)	2	.12	.36	.48
	3	.24	.09	.33
		.43	.57	1.00

be slicing the table the other way, lifting out a row and standardizing the probabilities.

Suppose that we randomly select a center director out of the respondents from the survey. Without knowing whether the director is from an urban or rural setting, the probability that the director's rating is 1, 2, or 3 is given by p(y). If we are then told that the director is from an urban setting, X = 1, we can revise our probabilities concerning the director's responses from the unconditional distribution p(y) to the conditional distribution p(y|x) where X = 1. In this simple example, we were able to obtain p(y|x) from p(x, y) and p(x) using equation 4.9. In the statistical inference problems to be discussed in subsequent chapters, the objective will be to revise a prior distribution p(y), on the basis of information about X, to the posterior conditional distribution p(y|x). However, the structure of the problem will be such that we will only know p(y) and p(x|y); the other distributions will not be available directly. Looking at Bayes' theorem 4.10, we can see that it is exactly what is needed for carrying out this revision; this will be the focus of much of the subsequent discussion.

THE CONTINUOUS CASE

In instances where the elementary events and the values of the random variable are continuous, the probability associated with any specific value of the random variable is zero. Thus rather than work with probabilities of particular values of random variables, we work with probabilities that random variables have values within certain intervals. The probability that the value of X lies in some interval with endpoints a and b can be obtained from the probability density function for X. We will denote this as p(x) because it is, in a general sense, analogous to the probability function in the discrete case.

The probability density function is such that

$$P(a \leqslant X \leqslant b) = \int_a^b p(x)\,dx. \qquad [4.12]$$

Suppose that we plotted a curve showing the values of the density function for each value of X. The probability density function has the property that the area under the curve between a and b represents the probability that the value of X lies in that interval. The integral in 4.12 simply represents the procedure for finding that area. The probability density function is constructed so that the area under the curve is unity; therefore, the probabilities associated with a set of exclusive and exhaustive intervals must add up to one. The value of the density function at some specific point x is not a probability and may be greater than 1.0. In the applications to be discussed in subsequent chapters, we will not need to work with these individual densities, but rather with probabilities pertaining to intervals.

In a manner analogous to equation 4.9 the conditional density function of Y given X is defined as

$$p(y \mid x) = \frac{p(y, x)}{p(x)}$$

where p(y, x) is the joint density function of Y and X, and $p(x) \neq 0$.

From this definition of conditional density functions, we obtain a version of Bayes' theorem for continuous random variables

$$p(y \mid x) = \frac{p(x \mid y)p(y)}{p(x)} \qquad [4.13]$$

If Y has values that range from $-\infty$ to $+\infty$, then the denominator can also be written

$$p(x) = \int_{-\infty}^{+\infty} p(x|y)p(y)\,dy$$

The denominator here serves the same purpose that it does in the discrete case: it ensures that the posterior distribution adds (in this case, integrates) to one. The symbol "p" is used to denote both probability functions in the discrete case and probability density functions in the continuous case so that we can easily discuss general features of Bayesian statistics that apply in both cases without having to refer to two different sets of terms.

BAYES' THEOREM FOR PARAMETERS AND OBSERVATIONS

We now turn to the use of Bayes' theorem for parametric statistical inference. We will see how the theorem can be used in making inferences about population parameters conditional upon a sample of observations. Before we can do this, however, we need to consider Bayes' theorem for vector variables.

Up to this point we have discussed the use of Bayes' theorem where we have a single observation or piece of information, and where we are interested in the prior and posterior distribution of a single variable. In many applications, we will be interested in revision of belief given a number of observations. Although one could use the theorem repeatedly with each observation, this would be too cumbersome for many problems, and we will want to consider the impact of a number of observations simultaneously. Furthermore, in many statistical applications we will need to consider prior and posterior distributions that involve more than a single variable.

Suppose that we have n observations $x_1, x_2, \ldots, x_n$, and k unknown quantities $y_1, y_2, \ldots, y_k$ about which we would like to revise our probabilities. We will represent the set of x's by **x** and the y's by **y**. **x** and **y** are *vectors*; they are made up of a number of quantities. Vectors are distinguished from *scalars*, which consist of a single quantity. Note that $p(\mathbf{x}) = p(x_1, x_2, \ldots, x_n)$; $p(\mathbf{x})$ represents a joint distribution. It is the distribution of probability over the possible combinations of values of the n observations. Likewise, $p(\mathbf{y})$ represents a joint distribution over the possible combinations of the k quantities.

Bayes' theorem holds for vector variables, as well as scalar variables

$$p(\mathbf{y}|\mathbf{x}) = \frac{p(\mathbf{x}|\mathbf{y})p(\mathbf{y})}{p(\mathbf{x})} \qquad [4.14]$$

This is like equations 4.9 and 4.13 except that the probabilities or densities are joint probabilities or joint densities. The basic operation is the same, however.

Bayes' theorem as expressed in 4.14 provides the basis for the Bayesian methods of statistical inference that will be discussed in the remainder of the book. Recall that the problem of parametric statistical inference is to make statements about unknown parameters that are not directly observable, from observable random variables the behavior of which is influenced by these parameters. The model of the data generation process specifies the relationship between the parameters and the observations. If $\mathbf{x}$ represents a vector of n observations $x_1, x_2, \ldots, x_n$, and Θ represents a vector of k parameters $\theta_1, \theta_2, \ldots, \theta_k$, on which the distribution of the observations depends, the model will be written $p(\mathbf{x}|\Theta)$. Writing 4.14 using Θ in place of $\mathbf{y}$, we have

$$p(\Theta|\mathbf{x}) = \frac{p(\mathbf{x}|\Theta)p(\Theta)}{p(\mathbf{x})}$$

When we are simply interested in making statements about the parameters Θ, given a particular set of observations $\mathbf{x}$, $p(\mathbf{x})$ is a constant that serves merely to make $p(\Theta|\mathbf{x})$ sum or integrate to one, as we have seen. Bayes' theorem can therefore be alternatively written as follows:

$$p(\Theta|\mathbf{x}) \propto p(\mathbf{x}|\Theta)p(\Theta) \qquad [4.15]$$

where $\propto$ indicates a relationship of proportionality. The constant of proportionality can always be obtained, if necessary, by using the fact that $p(\Theta|\mathbf{x})$ must sum or integrate to one.

In 4.15, $p(\Theta)$ is the joint distribution of the parameters prior to the observations becoming available, $p(x|\Theta)$ is the joint distribution of the observations conditional upon the values of the parameters, and $p(\Theta|x)$ is the joint distribution of the parameters posterior to the observations becoming available.

The "change agent" here, which serves to convert the prior distribution into a posterior distribution, is $p(x|\Theta)$. This is the model that relates the observations to the parameters. Before the observations are available, the model provides the joint distribution of the observations conditional upon the parameters. Once the sample is obtained, however, $p(x|\Theta)$ can be viewed as a function of Θ and is known as the *likelihood function* for Θ given x. That is, once x is given, the model can be used to determine the likelihood of obtaining those specific observations for different values of the parameters. The likelihood function is represented here by $\ell(\Theta|x)$. Bayes' theorem can be alternately written as

$$p(\Theta|x) \propto \ell(\Theta|x)p(\Theta) \qquad [4.16]$$

The likelihood term in Bayes' theorem expresses the information about the parameters provided by the observations. It is through the likelihood that the observations modify prior probabilities for the parameters. Given the prior distribution of the parameters and the likelihood function of the parameters given the observations, the posterior distribution of the parameters given the observations is determined.

As noted in Chapter 3, the mathematical derivation of Bayes' theorem is not controversial. Indeed, in the examples discussed in this chapter, persons holding different views about the interpretation of probability could perform the same calculations. It is when we begin to talk about inference for parameters using Bayes' theorem that the differences arise. We saw in Chapter 2 that in classical inference, the only probabilities involved are (or are based on) the probabilities of observations conditional upon a particular model of the population or data-generating process. These probabilities have meaning as long-run relative frequencies. However, for most problems the user of classical methods has no basis for assigning frequency-based probabilities to the values of the parameters. The parameters are considered unknown, but fixed. As a consequence $p(\Theta)$ has no meaning in frequency terms, and Bayes' theorem is of little use for inference. Classical inference is based solely on $p(x|\Theta)$ or in some instances, $\ell(\Theta|x)$.

For those holding logical or subjective views, probability is a degree of belief, and $p(\Theta)$ simply represents belief about the value of the parameters. (In a sense, a parameter is seen as a random variable; however, because $p(\Theta)$ is a distribution of belief, Θ is often referred to as an uncertain quantity rather than a random variable.) In this framework, Bayes' theorem is a fundamental tool for revising belief about unknown parameters given new information. Because persons holding logical or subjective views can make extensive use of Bayes' theorem in statistical inference, they are termed "Bayesians." This is to some extent a misnomer and has sometimes led to the erroneous impression that the debate between classical and Bayesian statisticians is over Bayes' theorem, when in reality the debate stems from different interpretations of probability and their implications.

In classical inference, the solution to the problem takes the form of a statement about what would happen over the long run, if the experiment were repeated over and over indefinitely. In Bayesian inference, the solution to the problem takes the form of a revised distribution of belief for Θ. To the Bayesian, the posterior distribution is the full inferential solution to the problem. Thus inference from data is basically revision of belief. Jackson (1976: 3) writes that Bayes' theorem is "the mathematical analogue of, or indeed the mathematical formulation of, the process of learning by experience." In the following chapters, we will look at how equation 4.16 is applied in specific problems.

FURTHER READING

Bayes' theorem is mentioned in most probability and statistics texts, although only the elementary forms are discussed in non-Bayesian presentations. Material on the same level as that presented here can be found in Novick and Jackson (1974), Phillips (1973), Schmitt (1969), and Winkler (1972). Weinstein and Fineberg (1980) describe the use of Bayes' theorem in diagnostic problems and Hunter (1984) discusses political and military applications. On the lighter side, see "There's No Theorem like Bayes' Theorem" set to the tune of "There's No Business like Show Business" by G.E.P. Box in Bernardo et al. (1980: 12-14).

5

Principles of
Bayesian Inference

In this chapter we examine some general features of Bayesian inference and issues surrounding its use, and we introduce some basic elements of inference for the normal model. The discussion will center on inference for normal means, in the case where the variance is known. In practice, there are few evaluation studies where we will know the true value of the variance for the process or population of interest. However, this case provides a useful introduction to the normal model that underlies many of the analyses discussed in this book. Furthermore, inference for this model involves many of the same considerations as inference for more realistic, but more complex, models and a number of important points can be illustrated with a minimum of distributional and notational complexity. More complex problems will be discussed in detail in subsequent chapters. At this point, we will be primarily concerned with the problem of inference—providing a probability statement in the form of a posterior distribution about the unknown quantity of interest. Decision making, which entails combining this inferential result with utility or loss considerations in making a choice among actions, will be considered in some detail in Chapter 9. We will first consider the basic theory, and then discuss a number of issues surrounding the use and specification of prior distributions. Following this an example is presented in which Bayesian and classical inferences are compared.

INFERENCE FOR A NORMAL MEAN
WITH KNOWN VARIANCE

THE NORMAL MODEL

Consider the case where the probability model for some random variable X is normal with unknown mean θ and known variance σ^2. The mean is denoted by θ rather than μ to emphasize that it will be treated as a random variable rather than as a fixed, but unknown, constant. Because the variance is known, it will continue to be represented by σ^2 in this chapter. The probability density function of X, given the parameters θ and σ^2, is given by the formula

$$p(x|\theta, \sigma^2) = (2\pi\sigma^2)^{-\frac{1}{2}} \exp\left[\frac{(x-\theta)^2}{-2\sigma^2}\right]$$

where "exp" represents the constant e = 2.718 raised to the power of the term in brackets. Given values for the parameters θ and σ^2, the density is fully determined for any particular value X = x.

In the problems that will be considered, there will usually be more than just a single observation. If we obtain n independent observations $\mathbf{x} = (x_1, x_2, \ldots, x_n)$ from a normal process with mean θ and variance σ^2, the joint probability density of the observations is the product of the individual densities:

$$p(x|\theta, \sigma^2) = \prod_{i=1}^{n} p(x_i|\theta, \sigma^2)$$

where the symbol Π represents a product operator. Carrying out this multiplication and doing some rearranging results in

$$p(\mathbf{x}|\theta, \sigma^2) = (2\pi\sigma^2)^{-n/2} \exp\left[\frac{\sum\limits_{i=1}^{n}(x_i - \bar{x})^2 + n(\bar{x} - \theta)^2}{-2\sigma^2}\right] \qquad [5.1]$$

where $\bar{x}$ is the mean of the n observations.

THE LIKELIHOOD FUNCTION

As discussed in the previous chapter, once the data are obtained, equation 5.1 can be viewed as a function of θ. For any value of θ, we can calculate the probability density for the data that were obtained. This is the likelihood function $\ell(\boldsymbol{\Theta}|\mathbf{x})$. It is important to note that the likelihood function is not a probability function; it does not sum or integrate to one. A probability function expresses the probability or density of the possible observations for a single model; the likelihood function, on the other hand, expresses the probability or density of a particular set of observations for many models with different values of the parameter θ. There is no reason that these values for different models should sum or integrate to one. In fact, the likelihood function is defined only up to a multiplicative constant. Any function that is a multiple of $\ell(\theta|\mathbf{x})$ is also a likelihood function for θ given $\mathbf{x}$. We therefore write

$$\ell(\theta|\mathbf{x}) \propto (2\pi\sigma^2)^{-n/2} \exp\left[\frac{\sum\limits_{i=1}^{n}(x_i - \bar{x})^2 + n(\bar{x} - \theta)^2}{-2\sigma^2}\right] \qquad [5.2]$$

The relationship here is one of proportionality.

We can see that with a fixed number of observations and known variance, the term $(2\pi\sigma^2)^{-n/2}$ will simply be a multiplicative constant. Notice also that the exponent is composed of two terms:

$$\sum_{i=1}^{n}(x_i - \bar{x})^2 / -2\sigma^2$$

and $n(\bar{x} - \theta)^2/-2\sigma^2$. The first term involves only fixed quantities—the observations that were obtained and the known variance—and this too can be factored out of the exponent as a multiplicative constant. Because the relationship in expression 5.2 is proportional, these multiplicative constant terms can be dropped and we simply write

$$\ell(\theta \mid x) \propto \exp\left[\frac{n(\bar{x} - \theta)^2}{-2\sigma^2}\right]$$

The remaining term is called the *kernel* of the likelihood and the deleted portion is called the *residue*.

A look at this kernel reveals that the observations enter the likelihood function only through the statistics $\bar{x}$ and n. For the normal model with known variance, $\bar{x}$ and n are said to be *sufficient statistics*. From the Bayesian viewpoint these statistics are sufficient in the sense that they lead to the same posterior distribution as would be obtained using the full set of observations. Sufficient statistics have some value here for "boiling down the data." It may not always be possible to find a small set of sufficient statistics for all models; however, when they do exist, they help to simplify Bayesian inference. A formal definition and discussion of sufficient statistics in the Bayesian framework can be found in Raiffa and Schlaifer (1961).

THE PRIOR AND POSTERIOR DISTRIBUTIONS

Consider the case where prior belief about the unknown mean θ is concentrated around some value and trails off symmetrically on either side. It may be possible to fit this prior distribution with a member of the normal family. Fitting procedures will be considered in more detail later; for now we will look at the implications of this prior. This prior distribution will be represented by $\theta \sim N(\mu_0, \sigma_0^2)$; that is, the prior probability distribution of θ is normal with mean μ_0 and variance σ_0^2. (Use of the symbols μ_0 and σ_0^2 is unrelated to their use as null values in Chapter 2.) Keep in mind that we are discussing two kinds of parameters here. θ and σ^2 are parameters of the model describing the population, or data-generating process. The mean θ is unknown, and μ_0 and σ_0^2 are parameters of the prior distribution of subjective probability over the possible values of θ. In other words, the model parameter θ is viewed as having a distribution that in turn has parameters. These second-level parameters are called *hyperparameters*. They will continue to be referred to occasionally as parameters when the meaning is clear from the context.

The prior density function of θ, which will be represented by $p(\theta)$, is written

$$p(\theta) = (2\pi\sigma_0^2)^{-\frac{1}{2}} \exp\left[\frac{(\theta - \mu_0)^2}{-2\sigma_0^2}\right] \qquad [5.3]$$

where μ_0 and σ_0^2 are considered to be fixed. This prior distribution can also be broken down into a kernel that is the exponential term, and a constant residue $(2\pi\sigma_0^2)^{-\frac{1}{2}}$. Because the posterior distribution is proportional to the product of the prior distribution and the likelihood, as in expression 4.16, the residue of the prior can be dropped from the present calculations without altering the relationship of proportionality. The posterior distribution of θ given the sample sufficient statistics $\bar{x}$ and n is the product of the prior and the likelihood and is obtained as follows:

$$p(\theta \mid x, n) \propto \exp\left[\frac{(\theta - \mu_0)^2}{-2\sigma_0^2}\right] \exp\left[\frac{n(\bar{x} - \theta)^2}{-2\sigma^2}\right] \qquad [5.4]$$

With some algebraic manipulation, the product 5.4 reduces to

$$p(\theta \mid \bar{x}, n) \propto \exp\left[\frac{(\theta - \mu_n)^2}{-2\sigma_n^2}\right] \qquad [5.5]$$

where

$$\mu_n = \frac{(\sigma_0^2)^{-1} \mu_0 + (n^{-1}\sigma^2)^{-1} \bar{x}}{(\sigma_0^2)^{-1} + (n^{-1}\sigma^2)^{-1}} \qquad [5.6]$$

and

$$\sigma_n^2 = [(\sigma_0^2)^{-1} + (n^{-1}\sigma^2)^{-1}]^{-1} \qquad [5.7]$$

Expression 5.5 is a kernel of a normal distribution; it has the same form as the kernel of the prior in expression 5.3 except that the mean has been revised from μ_0 to μ_n, and the variance has been revised from σ_0^2 to σ_n^2 on the basis of the data. Thus the posterior distribution of θ is normal with a mean of μ_n and a variance of σ_n^2; we write a posteriori, $\theta \sim N(\mu_n, \sigma_n^2)$. Again μ_n and σ_n^2 are hyperparameters, or are parameters that index the subjective probability distribution of the unknown model parameter—the mean θ. Should we need the residue for the posterior distribution, the fact that it must integrate to one requires that this residue be $(2\pi\sigma_n^2)^{-\frac{1}{2}}$.

The important point here is that, if the data are generated by a normal process with known variance, and if prior belief about the unknown mean θ can be represented by a normal distribution, then the posterior distribution of θ will also be normal, and the hyperparameters of this posterior distribution can be obtained easily from the prior hyperparameters and the sample sufficient statistics using formulas 5.6 and 5.7.

NATURAL CONJUGATE PRIORS

The normal prior distribution is said to be a *natural conjugate* to the likelihood function for the normal model with known variance. The family of normal distributions is a *natural conjugate family* for this likelihood function. This family is said to be *closed under sampling* in that a prior from this family will always combine with data generated by a normal process with known variance to yield a posterior distribution in the same family.

Raiffa and Schlaifer (1961: 43-44) discuss certain properties that are desirable in a family of prior distributions. First, a prior distribution from this family should be analytically tractable in three respects: It should be relatively easy to determine the posterior from the prior and the sample; the prior should lead to a posterior that is a member of the same family to facilitate repeated use of Bayes' theorem as the data accumulate; and it should be convenient to compute expected utilities for simple utility functions when using the posterior distribution in decision making. As we just saw in our discussion of the normal prior, the actual calculation of the posterior is simply a matter of entering in the appropriate prior and sample values in formulas 5.6 and 5.7. The normal family is closed under sampling—the normal prior leads to a normal posterior. The normal prior can also be easily used in computing expectations in decision making; this topic will be considered in Chapter 9.

Second, the prior family should be rich in the sense that there is enough variety in the shape of the members of the family to enable one to represent various prior beliefs. In other words, by varying the parameters of the distribution, we should be able to find a reasonable approximation to prior beliefs. The normal distribution does assume a variety of unimodal shapes as the variance changes. It is true that prior belief may never correspond exactly with a member of some general family of distributions, just as any real-world data-generating process may not correspond exactly with the model chosen by the data analyst. What is required is that some member of the prior family provide a reasonable approximation to prior beliefs. We shall see that with increasing sample size, Bayesian inference becomes increasingly insensitive to variations in the prior.

Third, the parameters of the family of distributions should be interpretable to the person whose prior beliefs are being modeled so that it is easy to verify that the distribution chosen actually represents the individual's prior beliefs. In this case, the mean and the variance (or standard deviation) give some feel for the location and dispersion of the distribution. As we will see, examination of the role of prior parameters and the sample statistics in formulas 5.6 and 5.7 provides some feel for the relative impact of prior information in the revision process.

The normal family of conjugate priors has these properties for problems involving the normal process with known variance. The natural conjugate families for other models that will be considered also share these properties. The existence of sufficient statistics ensures that conjugate distributions can be found; however, Bayesian analysis does not depend upon the existence of sufficient statistics and natural conjugate distributions—a posterior distribution can always be obtained through numerical methods. However, this can involve considerable computational effort. All the models discussed in this book have natural conjugate families that have the properties just discussed, and the discussion will be limited to natural conjugate priors and posteriors.

THE REVISION PROCESS

Let's take a closer look at what happens as the prior distribution is revised. Referring back to formula 5.6 reveals that the posterior mean μ_n is simply a weighted average of the prior mean μ_0 and the sample mean $\bar{x}$ with each mean weighted by the reciprocal of its associated variance. In formula 5.7 we can see that the reciprocals of these variances enter into the posterior variance also. The posterior variance is simply an inverse of their sum.

Prior and Posterior Precision

The meaning of these formulas can be seen more clearly if the precision of a normal distribution is defined as the reciprocal of its variance. The smaller the variance the greater the precision. A precise distribution is highly concentrated around its mean. The prior precision is $h_0 = (\sigma_0^2)^{-1}$, the sample precision is $h = (n^{-1}\sigma^2)^{-1}$, and the posterior precision is $h_n = (\sigma_n^2)^{-1}$. We can then write the formula for the posterior mean as

$$\mu_n = \frac{h_0\mu_0 + h\bar{x}}{h_0 + h} \qquad [5.8]$$

and we can write the formula for the posterior precision as

$$h_n = h_0 + h \qquad [5.9]$$

It can be seen in formula 5.8 that the posterior mean μ_n is a weighted average of the prior mean μ_0 and the sample mean $\bar{x}$, each weighted by the appropriate precision, h_0 and h, respectively. Two points regarding the relative impact of prior and sample information are worth noting here. First, from the definition of the sample precision, we see that as the sample size increases, the sample precision increases. Therefore, as the amount of data increases, the effect of the sample on the posterior distribution increases, and beliefs about θ are increasingly brought into line with the data. Looking at this in another way, as the data accumulate, the effect of the prior distribution on the posterior distribution becomes less. Second, the variance of the prior distribution reflects the amount of prior uncertainty about the value of θ. With a large variance, the prior distribution is spread out over a large range of values of θ, thus expressing substantial uncertainty about this quantity. When the prior variance becomes larger, the prior distribution becomes increasingly flat or uniform, and a prior distribution approaching uniformity is sometimes seen as a way of representing a situation where prior information about θ is vague. When the prior is flat, no value of θ is more likely than any other. As the prior variance becomes increasingly large, the prior precision approaches zero and the posterior distribution is determined almost entirely by the data, as would seem reasonable when there is little prior information. Indeed, when the prior precision goes to zero, the posterior mean of θ equals the sample mean. This type of "noninformative" prior can be quite useful in Bayesian analysis.

It can be seen in formula 5.9 that the posterior precision is simply a sum of the prior and sample precisions. Here again the relative impact of the prior and the sample depends upon their precision. As more and more data are obtained, the sample precision and the subsequent posterior precision become greater. This increased precision expresses increased certainty about the value of θ. Again, when the prior is imprecise, the posterior precision is determined largely by the sample. When the prior precision goes to zero, the posterior precision equals sample precision.

An Equivalent Prior Sample

We can look at the combination of prior and sample in another way. Suppose that before obtaining the present sample with sufficient statistics $\bar{x}$ and n, we had obtained a previous sample of m observations with a mean of $\bar{w}$ from the same normal process. If we used a vague or noninformative prior in analyzing this previous sample, the posterior distribution would be $\theta \sim N(\bar{w}, m^{-1}\sigma^2)$. That is, with a noninformative prior the posterior mean of θ is the same as the sample mean $\bar{w}$, and the posterior variance of θ is the same as the sample variance $m^{-1}\sigma^2$.

Now suppose that we use this as a prior distribution in analyzing our present sample. The posterior mean of θ is then

$$\mu_n = \frac{(m^{-1}\sigma^2)^{-1}\,\bar{w} + (n^{-1}\sigma^2)^{-1}\,\bar{x}}{(m^{-1}\sigma^2)^{-1} + (n^{-1}\sigma^2)^{-1}} = \frac{m\bar{w} + n\bar{x}}{m + n}$$

and the posterior variance of θ is

$$\sigma_n^2 = [(m^{-1}\sigma^2)^{-1} + (n^{-1}\sigma^2)^{-1}]^{-1} = (m+n)^{-1}\sigma^2$$

We see that the relative impact of these two samples depends only upon the number of observations m and n. Note that these formulas simply involve pooling of the data from the two samples. In fact, the posterior hyperparameters will be the same whether the data were

TABLE 5.1
Inference for the Mean of a Normal Distribution
(σ^2 known)

(a)   Using a Noninformative Prior Distribution

Prior Distribution:
$\theta \sim$ uniform

Sample Sufficient Statistics:
$n, \bar{x}$

Posterior Distribution:
$\theta \sim N(\bar{x}, \; n^{-1}\sigma^2)$

(b)   Using an Informative Prior Distribution

Prior Distribution:
$\theta \sim N(\mu_0, \; \sigma_0^2)$
or
$\theta \sim N(\bar{w}, \; m^{-1}\sigma^2)$

Sample Sufficient Statistics:
$n, \bar{x}$

Posterior Distribution:

$$\theta \sim N\left[\frac{(\sigma_0^2)^{-1}\mu_0 + (n^{-1}\sigma^2)^{-1}\bar{x}}{(\sigma_0^2)^{-1} + (n^{-1}\sigma^2)^{-1}}, \; [(\sigma_0^2)^{-1} + (n^{-1}\sigma^2)^{-1}]^{-1} \right]$$

or

$$\theta \sim N\left[\frac{m\bar{w} + n\bar{x}}{m + n}, \; (m+n)^{-1}\sigma^2 \right]$$

obtained in two separate samples or one combined sample of $m + n$ observations.

Even if the prior distribution is not based on a previous sample we can still express the prior parameters μ_0 and σ_0^2 in these terms. If we let $m = \sigma^2/\sigma_0^2$, then $\sigma_0^2 = m^{-1}\sigma^2$. Now if we let $\mu_0 = \bar{w}$, we can think of our prior information as equivalent to having obtained a previous sample of m observations with a mean $\bar{w}$ and a variance $m^{-1}\sigma^2$. We could say that our prior beliefs are "worth" m observations. The relative impact of the equivalent prior sample and the actual sample depends upon m and n. Looking at prior information in this way can be useful in thinking about and assessing prior distributions.

Some of these results are summarized in Table 5.1 for reference purposes. The top portion of the table shows the prior, the sample sufficient statistics, and the posterior for inference about the unknown mean σ with noninformative prior. The bottom portion of the table shows the same elements for inference with an informative prior. The prior and posterior distributions are expressed in terms of μ_0 and σ_0^2, and in terms of the equivalent prior sample form just discussed.

In summary, Bayes' theorem and the derivations from it specify how prior and sample information should be combined in the revision of belief. The posterior distribution describes what one should believe about the unknown parameter given the new data. To the Bayesian, the posterior distribution is the complete inferential statement about the unknown parameter. Given the posterior distribution, one can make a variety of probabilistic statements about the unknown parameter. In the case just discussed, the posterior distribution is normal with mean μ_n and variance σ_n^2. Thus $z = (\theta - \mu_n)/\sigma_n$ has a standard normal distribution with a mean of zero and a standard deviation of one. By referring to a cumulative distribution for a standard normal variate, as tabulated in most statistics texts, we can determine the probability that θ is greater than or less than any particular value of interest, or that θ lies in some interval of interest. These kinds of inferences will be considered in subsequent examples.

Note that the posterior distribution is an inferential statement about the unknown parameter—it simply expresses the probability that θ takes on certain values. By itself, the posterior distribution does not dictate any particular action. Any decision that depended upon θ would require consideration of loss or utility, in addition to the probabilities expressed in the posterior distribution. The important point for the use of evaluation findings is that the Bayesian approach spells out what one should believe on the basis of the data. The classical approach imposes no order on any revision process; this is left to the informal judgment of the user.

THE PRIOR DISTRIBUTION

The prior distribution is a distinguishing feature of Bayesian statistics. Without a prior distribution there is nothing to be revised via Bayes' theorem. In some instances where prior information is considered vague in relation to that to be obtained from the sample of observations, one may simply carry out this analysis using a noninformative prior. In other instances, one may want to specify prior belief in some detail and incorporate this in the analysis. This section will consider certain points relating to the specification of prior distributions and their use. However, we first consider some general issues relating to objections to the use of prior distributions in principle, whatever their form and degree of specification.

SOME GENERAL ISSUES

Some critics might argue that they have no prior probabilities that they could specify concerning parameters, and the Bayesian approach is, therefore, of no value to them. As a response to this, it should be recalled that the coherence argument says that if you behave in such a way as to satisfy the axioms of coherence, then you are behaving as if you had a prior distribution. That is, your behavior is the same as if you were a Bayesian who specified a prior distribution. If you are not behaving as if you were a Bayesian, then some axioms are being violated. It would seem unreasonable to behave in a way that violates these simple axioms or to behave in a way that implies the existence of a prior that does not reflect your true beliefs. Therefore, the most straightforward approach would be to give some serious thought to one's prior beliefs and introduce them explicitly at the outset. Although this argument does not suggest an easy way to specify prior beliefs, it does mean, more generally, as Meier (1981: 342) states, "that if we have methods of statistical inference that are *inconsistent* with *any* Bayesian framework, they must somewhere lead us to incoherent decision making and that they presumably are not good general methods."

A second concern that is sometimes expressed is that even if we do have prior beliefs, they are difficult to measure exactly and, consequently, they should not be formally introduced into statistical analysis. It needs to be kept in mind, however, that while a particular expression of prior belief may not be an exact description, neither is the model of the data-generating process likely to be an exact description of that process. The point is that the prior distribution, like the model (and the derived sampling distribution used in the classical analysis), is an assumption that enables us to proceed. Leamer (1978: 41) writes,

A word about assumptions is in order. Anyone who insists that personal probabilities be assessed precisely cannot perform statistical inference for the same reason that no spaceship would ever have reached the moon if measurement of lengths had to be perfectly precise before construction could commence. To say something is 6.5 centimeters long is only to say that for the purposes at hand we may proceed as if it were. An assumption is not, therefore, a statement of unquestioned truth. It is a tentative statement on which initial action can usefully be based.

It is interesting that persons who are willing to assume or make the judgment that a data-generating process for some variable follows the formula for the normal function described earlier in this chapter, without giving it a great deal of thought, express considerable concern over the exact modeling and measurement of prior belief. Precise modeling of the data-generating process may, of course, be unnecessary in many problems because the inferences are relatively robust with respect to deviations from the actual process. Similarly, we will see that for many Bayesian analyses, the inferences will be robust with respect to measurement error in the prior. Indeed, it is difficult to talk about quality of measurement in the abstract. For some problems, exact measurement may be necessary; for other problems only approximate measurement may be needed. The sensitivity of one's conclusions and decisions to variations in the prior distribution is most usefully considered in the context of a specific problem. One generalization, however, is that as the sample size increases, the exact form of the prior distribution becomes less and less critical. With large sample sizes, the resulting inferences will be virtually the same over a wide range of prior distributions.

In some situations, precise specification may be important, and it may require some effort. Berger (1980: 84) comments in this regard that "priors are hard to construct. If they are needed for a sound analysis of the problem, however, this is merely an unfortunate fact of life." Sticking one's head in the sand and refusing to consider prior information because it is hard to measure is not a solution to the problem. Winkler (1982a: 243) writes that although it may not always be easy to apply Bayesian methods, this can be viewed as a positive feature rather than as a weakness—it forces the statistician to do some hard thinking about the modeling of prior belief, which should affect the conclusions that will be reached.

A third concern is that incoporating prior belief into the analysis introduces a subjective element into what should be a purely objective process. If by this it is meant that personal judgment should be eliminated from the analysis, then it needs to be recognized that the analysis must involve judgment. Fairly specific judgments or prior beliefs about parameters are introduced into the classical analysis, as in the Bayesian analysis, in the form of specification and measurement assumptions concerning the following: variables to include in an analysis, homogeneity of variance, absence of interaction effects, uncorrelated error terms, and so on. (It is curious, as Box, 1980: 384, notes, that for persons critical of the use of priors, assumptions somehow do not count as prior belief.) In addition, prior belief is introduced informally in the interpretation of results. Consider the example set forth by Savage (1961) of the following three experiments (as described in Berger, 1980: 2):

(1) A lady, who adds milk to her tea, claims to be able to tell whether the tea or the milk was poured into the cup first. In all of ten trials conducted to test this, she correctly determines which was poured first.
(2) A music expert claims to be able to distinguish a page of Haydn score from a page of Mozart score. In ten trials conducted to test this, he makes a correct determination each time.
(3) A drunken friend says he can predict the outcome of a flip of a fair coin. In ten trials conducted to test this, he is correct each time.

In these experiments, the quantity or parameter of interest is the true probability of being able to give the correct response. The probability of obtaining 10 correct answers, if the true probability of answering correctly were .50, or mere chance, is $(.50)^{10}$. A one-sided test of the null hypotheses that the true probability is .50, would lead to a rejection of H_0 with $p < .001$ in each experiment. Yet our interpretation of the results is likely to differ considerably in the three experiments. In the second experiment, it seems reasonable to conclude that the true probability is greater than the chance level of .50. An expert should be able to make these determinations; this conclusion is consistent with our prior beliefs. In the third experiment, our prior probability that our friend has some prescient ability is likely to be very low, and most of us would be reluctant to conclude that he has such ability on the basis of the experiment. We would be more inclined to attribute the result to a rare streak of luck or to some alternative mechanism such as a party trick. In the first experiment, prior beliefs may vary from person to person and different people may interpret the outcome differently. Yet the null hypothesis is rejected at the same level in all three experiments. Prior belief does have an effect on what we conclude from the data. Failure to acknowledge this can lead to much misdirected debate. While individuals may not be operating with precisely quantified prior distributions, prior information is relevant to their interpretation of the results. Rather than leave this to unaided intuition, the Bayesian approach formalizes this with its explicit treatment of prior information, and makes its implications clear.

It is true that, in many problems, there will be substantial agreement on the appropriate model, while different individuals may have different priors. Consequently, they could reach different conclusions on the basis of the same data. The goal of scientific inference, some will argue, is not to produce such individualized conclusions, but to produce conclusions with which we would all agree. Indeed, the experimental method emphasizes drawing the conclusions from the data. It would seem, then, that only the data are relevant for reaching scientific conclusions. It can be argued, however, that this is a misleading extrapolation from controlled experimentation in the physical sciences (Novick and Jackson, 1974: 145-146; Jackson, 1976: 7-10). They point out that in the physical sciences, experiments are carried out in carefully controlled environments and are repeated until sufficient evidence is available to reach a firm conclusion. In terms of our earlier discussion, this has the effect of decreasing the error variance so that the sample becomes sufficiently precise as to make the effects of the prior negligible. Novick and Jackson (1974: 146) write that "what characterizes scientific results is not 'objectivity' but sample precision." In the field of evaluation, and in the social sciences in general, it is not always possible to conduct experiments in tightly controlled settings. With diminished sample precision, prior information may be expected to have some bearing on the conclusions reached. Ignoring prior information does not make research more scientific. It is interesting to speculate that much of the disagreement over the conclusions of evaluation studies may stem from unacknowledged differences in prior opinion.

As we have argued, prior beliefs do and should affect statistical analysis. As De Finetti (1974b: 128) points out, refusing to spell out prior probabilities does not make the analysis objective—the analysis will either be untenable whatever the prior might be, or equivalent to blind adoption of some prior. In fact, the Bayesian approach could be said to be more objective than the classical approach in this respect because it does enable one to spell out prior belief and makes its implications explicit. If there is prior information available, it could hardly be called objective to ignore its implications for the analysis.

ASSESSMENT OF PRIOR DISTRIBUTIONS

There are a number of different approaches for assessing prior distributions depending upon the nature of the prior information available.

Data-Based Priors

The most straightforward case is that in which one's prior information consists of previous data on the unknown quantity. These data can be used to derive what have been called *data-based priors* (Zellner, 1971a) or *empirical priors* (Mosteller and Tukey, 1968). In certain situations, we may have access to historical data on the value of the parameter. These previous values could be considered as observations from a prior distribution of the unknown parameter and used to determine its form. With a sufficient number of previous values, one could construct a frequency distribution, and with some smoothing use this as one's prior.

Unfortunately, for many evaluation problems, previous values of the unknown parameter of interest will not be available. We may, however, encounter the situation in which a previous sample has been obtained under similar circumstances. If this sample were analyzed using Bayesian methods, we could then use the posterior distribution obtained from this previous sample as a prior distribution in analyzing a subsequent sample. This still leaves us in the position of needing a prior distribution to analyze the previous sample. As was discussed above, the initial prior has less and less effect on the posterior as the data accumulate and, within certain broad limits, the exact form of this prior will make little difference. There is justification, then, for beginning the whole process with some vague or noninformative prior to get the revision process started; we discuss this in more detail below.

Non-Data-Based Priors

Often, prior information derives from sources other than previous samples. In order to incorporate this information in a Bayesian analysis, the individual's beliefs must be quantified in terms of subjective probability. This type or prior is known as a *non-data-based prior* (Zellner, 1971a) or *belief prior* (Mosteller and Tukey, 1968).

In the subjective interpretation of probability, subjective probability for an event can be operationally defined in terms of the individual's preference between lotteries. In one lottery, the outcomes depend upon the occurrence or nonoccurrence of the event of interest. In the other lottery, involving the same outcomes, the outcomes depend upon the occurrence or nonoccurrence of some reference event with a known reference chance or probability for that individual. Finding the reference probability that makes the individual indifferent between the two lotteries provides the basis for scaling the individual's subjective probability for the event. This is known as the "indirect" approach. See Pratt et al. (1964, 1965) for a thorough discussion of the logic of this approach. Sometimes, subjective probabilities are obtained using a direct method of assessment. This involves simply asking the individual how likely he or she feels the occurrence of the events to be. This might involve asking for a probability, or it could involve asking for odds—the probability of one event relative to another. The requested responses might not even be probabilities. Respondents could simply be asked to divide up 100 poker chips over the events to represent how likely they are believed to be. The proportion of chips assigned to an event could be used as a measure of subjective probability. Certain checks for coherence could be included.

In this book, the concern is not with assessment of probabilities for events but rather for unknown parameters in the form of random variables. When the unknown parameter is discrete, the assessment task is essentially the same as that for events. When the parameter is continuous, the task is somewhat different. Because in this case there are an infinite number of possible values of the parameter, it is no longer a simple matter of asking for probability assessments for each value. Here also, there are a number of different methods for eliciting subjective probabilities. A simple direct approach is to divide up the range of the parameter into a number of intervals and obtain the subjective probability that it lies in that interval. This can be used to construct a histogram, which with some smoothing can be used as a prior distribution. The main disadvantage of this approach is that the resulting distribution may be difficult to work with. The density function may not be easily expressed in any analytically tractable form, and numerical methods may be necessary for the application of Bayes' theorem.

This book will focus on methods for matching a member of the natural conjugate family. As noted earlier, the advantage of using a member of the natural conjugate family is its computational simplicity. If the prior is conjugate to the likelihood function, then the posterior will be a member of the same family, and its hyperparameters will be obtained from the hyperparameters of the prior and the sample sufficient statistics by relatively simple algebraic methods. The natural conjugate families to be considered in this book are sufficiently rich that a variety of priors can be fit with members of the family. In addition, it will be seen that even if the fit is not exact, in many situations use of a member of the natural conjugate family will yield a posterior that is essentially the same as the posterior that would have otherwise been obtained.

Here also there are a number of different approaches. Two of the more commonly used methods involve use of fractiles and use of equivalent prior samples. Fractiles divide a distribution into areas containing a specific percentage or proportion of the total area in the distribution. For example, 50% of the distribution lies below the 50% fractile, and 75% lies below the 75% fractile, and so on. These fractiles can be interpreted in terms of probability—the probability that the random variable has a value less than the 50% fractile is .50, and so on. If an individual can specify a few points such that the probability of the unknown parameter being less than these points is some amount .50, .75, .90, for example, the individual's beliefs can be fit with a member of a particular conjugate family.

An alternative assessment method involves equating prior information with a sample of a particular size. We saw earlier how a prior distribution could be expressed as equivalent to a previously obtained sample. By matching an individual's beliefs with the implications of various samples, the individual's beliefs can be fit with a member of a particular conjugate family. Again, various coherence checks could be applied. These techniques are combined in the computerized assessment procedures that will be discussed in a following chapter.

The argument here is not that all priors can be fit by a member of an appropriate conjugate family, or that they should be forced into such form. The priors discussed in this book are sufficiently rich so as to enable one to approximate a variety of priors by a member of the natural conjugate family. However, Bayesian methods can still be employed when the prior is not in the form of a natural conjugate distribution. In this case, more complex numerical methods may be necessary to obtain the posterior distribution, and this can involve considerable computational effort. The discussion will be limited to natural conjugate priors and posteriors.

Vague Prior Information

When we have little prior information we may find it convenient to use a noninformative prior distribution—one that contributes little to the posterior distribution. In many problems where the unknown parameter ranges from $-\infty$ to $+\infty$, such as a normal mean does, a non-informative distribution is represented by one that is uniform. This uniformity reflects the lack of information concerning the values of the parameters. In looking at Bayes' theorem in expression 4.16, it can be seen that if the prior distribution $p(\Theta)$ is constant, then the posterior distribution is determined entirely by the likelihood. In other words, the results of the analysis depend entirely on the observations. Although this has a commonsense appeal, theoretical difficulties arise in finding probability distributions that are completely noninformative as perfectly uniform distributions ranging from $-\infty$ to $+\infty$ cannot sum or integrate to one as required by the axioms of probability theory and are improper distributions. This is generally not a serious difficulty because a workable approximation requires only that the distribution be locally uniform in the region where the likelihood has any appreciable value (Box and Tiao, 1973: 21). It is important not to misinterpret this noninformative prior as expressing total ignorance concerning the unknown quantity or parameter because to do so leads to contradictions. If we are completely ignorant about the value of θ, then we should also be completely ignorant about θ^2, or θ^{10}, or θ^{-10}, and so on. A uniform distribution on any one of these implies a nonuniform distribution on the others and there is no obvious reason to pick one over the other as being *the* expression of ignorance.

An individual will seldom, however, be in a state of total ignorance, and that will make certain of these parameterizations more acceptable than others to the individual as expressing his or her vague prior information. Consider an investigator whose prior beliefs are vague concerning how students will perform on a 100-point test that is being used as an outcome measure in an evaluation. The investigator feels that over a substantial range of possible scores, the group mean θ is likely to assume one value as another and considers $p(\theta)$ to be uniform over this range. In line with this, the investigator feels that θ is as likely to lie between 60 and 70 as it is to lie between 70 and 80. A uniform prior on θ^{10}, on the other hand, would imply that it is over three times as likely that θ would be between 70 and 80 than between 60 and 70, and this would be inconsistent with the investigator's beliefs. Again, it comes down to a judgment on the part of the individual. It is true that less extreme transformations, $\theta^{1.1}$ say, would be less inconsistent and the implications may be indistinguishable to the investigator. On the other hand, with even a moderate amount of data, taking $p(\theta^{1.1})$ to be uniform, instead of $p(\theta)$, will lead to virtually identical results.

For our purposes, a noninformative prior simply represents prior belief that is vague or imprecise in relation to the data, rather than total ignorance. The use of a uniform prior is an approximation procedure: It enables us to obtain a posterior distribution that is approximately the same as one that would be obtained with a thorough assessment of prior information. The justification of such an approximation lies in what has been called the principle of *stable estimation* (Edwards et al., 1963) or *precise measurement* (Savage, 1962). The basic idea is that when the sample is precise in relation to the prior, the prior will have little impact on the posterior. A variety of different priors will lead to approximately the same posterior distribution—the posterior distribution obtained using a noninformative prior will be approximately the same as the posterior obtained using any prior that is dominated by the likelihood. The principle of stable estimation is what makes Bayesian inference practical; as the sample becomes increasingly precise, the posterior becomes increasingly robust against measurement error in the prior.

The main advantage of the noninformative prior is its simplicity. It spares us of the effort required to make precise assessments of prior belief, when such prior belief will have little impact on the results. Because the posterior is determined almost entirely by the data, it also has some value in reporting to a wide audience. Furthermore, we shall see also that when noninformative priors are used, there are certain numerical similarities between Bayesian and classical results. Although this is certainly not a justification for using noninformative distributions, it is useful in comparing classical and Bayesian methods. It is important to keep in mind, however, that uniform distributions are approximations that depend upon the relative precision of the prior and sample. Relatively imprecise priors are also called vague priors, diffuse priors, informationless priors, and so on; we have used the term "noninformative prior" to express this. (Box and Tiao, 1973, use the term "noninformative prior" in a more specific sense.)

There are various methods for finding priors that have little impact on the posterior. In many instances, these methods lead to the same noninformative prior, but in others they do not. Generally, with moderate amounts of data these different noninformative priors lead to similar, if not identical, results. With small amounts of data they can lead to different results; however, in this situation prior belief will not be imprecise in relation to the data, and the use of a noninformative prior of any type is a questionable practice.

CHOICE OF A PRIOR DISTRIBUTION
FOR SCIENTIFIC REPORTING

The theory specifies how a prior distribution should be revised in light of the data, but it does not dictate whose prior distribution should be considered. This depends upon the purpose of the study and the intended audience. When an individual is conducting the analysis solely for his or her information, or use in decision making, only his or her prior distribution is relevant. The situation is changed when there are multiple potential users with possibly different priors, as there are likely to be for evaluation studies. Obviously, a posterior distribution that is obtained using one individual's prior is going to be of limited interest or value to a general audience. The issue is how to conduct the analysis in a way that will be of use to multiple users. This is often a source of misunderstanding concerning the use of Bayesian methods. Savage (1962: 98-99) writes, "Sometimes listeners to an exposition of Bayesian statistics get the misimpression that they are being urged to publish their own opinions as their analysis of an empirical study." The goal of reporting Bayesian results is to provide the user with the necessary information so that, given his or her prior distribution, the user can arrive at his or her own posterior distribution. Assuming that a user accepts the model upon which the analysis is based, reporting the likelihood function will enable him or her to do this. Any user could then determine how his or her beliefs should be modified by the data.

Although this is reasonable from a theoretical perspective, it is likely that many readers of research reports will not have the inclination, time, or training to conduct such individualized analyses. Fortunately, there are a number of strategies for choosing prior distributions such that posterior distributions are obtained that can be useful to a general audience. These are described by Roberts (1977) and also by Jackson (1976). First, multiple analyses could be conducted using different prior distributions. The audience could then see the extent to which particular priors are modified by the data. This would be especially useful if there were clearly identified conflicting prior positions. It is possible that a variety of priors may lead to the same, or very similar, conclusions. Some graphical procedures for displaying the results for a variety of priors are described by Dickey (1973). Second, it might be possible to use data-based priors if generally accepted previous research findings are available. Third, the analysis could be conducted using a noninformative prior distribution. In this case, the posterior distribution would be determined entirely by the data. Box and Tiao (1973: 22) suggest using this prior as a reference prior; the investigator could thus say that "irrespective of what he or anyone else believed to begin with, the posterior distribution represented what someone who *a priori* knew very little about an unknown parameter should believe in light of the data." They also argue that, in addition to simply serving as a neutral standard, the uniform prior may approximate prior belief in many instances. They contend that scientific investigations are not usually carried out unless the information to be obtained is relatively more precise than the information available prior to the study, which means that the prior distribution can be represented as being locally uniform. Fourth, if the investigator believes that he or she can justify his or her prior distribution and convince others that it is appropriate, an analysis based on this prior distribution could be reported.

Depending upon the problem at hand, the investigator might want to use more than one of these strategies in reporting the results of a study. Whatever strategy is used, the investigator should, in addition to reporting the posterior distribution(s), describe and justify any prior distributions(s) and present the likelihood function. The data could then be reanalyzed using different prior distributions, if desired. Hildreth (1963) discusses the type of information that should be included in a report of a Bayesian analysis in order to maximize its usefulness for different users.

THE PROBLEM OF DISAGREEMENT

The fact that individuals with different priors might arrive at different posterior distributions on the basis of the same data is disturbing to some. If different individuals reach different conclusions from the same data, this would appear to limit the value of the Bayesian analysis as an aid to public debate and decision making. One response to this comes from Edwards et al. (1963: 201):

> If observations are precise, in a certain sense, relative to the prior distribution on which they bear, then the form and properties of the prior distribution have negligible influence on the posterior distribution. From a practical point of view, then, the untrammeled subjectivity of opinion about a parameter ceases to apply as soon as much data became available. More generally, two people with widely divergent prior opinions, but reasonably open minds will be forced to arbitrarily close agreement about future observations by a sufficient amount of data.

Mosteller and Wallace (1963: 302) make a similar point in discussing their results:

> Prior distributions are not of major importance. While choice of underlying constants (choice of prior distributions) matters, it doesn't matter very much, once one is in the neighborhood of a distribution suggested by a fair body of data. We conclude from this that the emphasis on the difficulty, even impossibility, of choosing prior distributions as a criticism of the use of Bayes' theorem is not well placed.

These arguments apply when there is a substantial amount of data available. It is possible, however, that the amount of data available at the time some conclusion is to be reached, or some action is to be taken, may not be sufficient to "swamp out" prior differences. In this instance, neither the Bayesian theory nor any other theory will magically produce agreement where none exists. Lindgren (1976: 379) writes that when individuals with different priors consider the data "either the prior is not crucial and they all come to the same conclusions anyway, or it is crucial and they *should* come to different conclusions. In the latter case, they would either have to gather more data or to resolve their differences, a process that is healthy in provoking thought and reevaluation." Prior information does, and should, affect the interpretation and credibility of research findings. It is generally a mistake to believe that a single study represents the only piece of information about the state of the world, and that conclusions and decisions should be based only on this information. Bayesian inference makes the impact of prior information explicit and points out differences that arise due to different priors. Most Bayesians would see this as an advantage of the theory. In discussing the use of Bayesian inference in decision making, Box and Tiao (1973: 20) write, "Far from nullifying the value of Bayesian analysis, the fact that such analysis shows to what extent different decisions may or may not be appropriate when different prior opinions are held, seems to enhance it. For problems of this kind any procedure which took no account of such opinion would seem *necessarily* ill conceived."

Although it would be possible for an individual to concoct a prior distribution after the fact that would yield a desired posterior distribution, anyone wanting to insist that a particular conclusion or decision is appropriate given his or her prior distribution would need to present this prior for scrutiny and explicitly defend it. As a tactical consideration, it must be recognized that if the analysis is being used in support of some position or action, one is unlikely to convince, or enlist the support of, skeptical others by showing that the conclusions from the analysis are determined largely by one's own prior beliefs. Either the prior distribution needs to be justified to such an extent that it is accepted by others or, as Novick and Jackson suggest, one needs to show that the conclusions are consistent with the prior beliefs of others.[1] Mosteller and Wallace (1963: 303) point out that this problem is not unique to Bayesian statistics:

> Some people ask "Who is to be the official guesser of priors?" and "Won't different people get different results?" The answers are "No one," and "Yes." The position is similar to that of the choice of the data distribution in any sort of statistical analysis. There is no official chooser of data distributions, and people who choose different ones do get different answers.

In the end, it comes down to the point that if we want others to accept our conclusions, it is necessary to convince them that our assumptions and methods are reasonable.

In practice, noninformative prior distributions are frequently used. In his review of Bayesian statistics, Roberts (1978: 15) writes, "What appears not to have happened is extensive reliance on non diffuse prior distributions, a reliance much feared by some early critics of Bayesian methods, who thought that posterior distributions would be manipulated at will by the insertion of an appropriate non diffuse prior." Many applications emphasize noninformative priors because of their value as reference priors, and as Box and Tiao argue, they may represent the actual state of prior information relative to the data to be obtained for many users of research findings. Additional motivation for the use of uniform priors no doubt comes from the fact that it is easier to assume a prior distribution to be uniform than it is to assess and defend one that is nonuniform. As long as the amount of prior information is small in relation to the amount of information in the sample, it will make little difference whether a noninformative prior or a more precisely specified prior is used; nevertheless, the user of the findings of Bayesian analyses should keep in mind that prior information is a legitimate part of Bayesian inference, and that techniques for its incorporation are available when this is desired. We will discuss applications involving both noninformative and informative priors in the following chapters.

A GRAPHICAL ILLUSTRATION

Some of the points made so far can be summarized in a graphical illustration involving various prior distributions. The problem concerns the number of days of hospitalization required for patient recovery following use of a new surgical procedure. Length of stay for patients receiving the new procedure is considered to be normally distributed with known variance $\sigma^2 = 5.29$ days, and a study of thirty randomly selected patients receiving the new procedure yields an average length of stay $\bar{x} = 14.82$ days.

Analysis of these data using a noninformative prior as outlined in Table 5.1 yields the posterior $\theta \sim N(14.82, 0.18)$. The prior, the standardized likelihood, and the posterior are shown in Figure 5.1. The prior and posterior are identified by the number 1 in brackets. The likelihood has been standardized so that it integrates to one. We see here that the posterior is the same shape as the standardized likelihood. This makes

PRIOR DISTRIBUTIONS

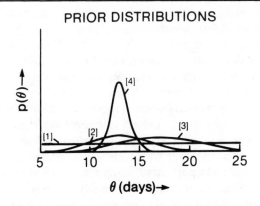

STANDARDIZED LIKELIHOOD

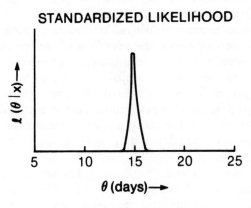

POSTERIOR DISTRIBUTIONS

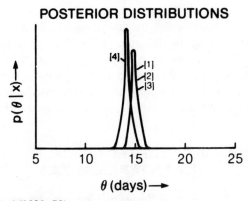

SOURCE: Pollard (1983: 73).

Figure 5.1 Posterior Distributions of Mean Days to Recovery θ Obtained Using 4 Different Prior Distributions

sense because in this case the posterior is entirely determined by the data as expressed in the likelihood.

Suppose that an individual believes, on the basis of some previous experience with surgical procedures of this type, that there is a 50% chance that θ will be 13 days or less, and that there is a 50% chance that θ will be between 11 and 15 days. This can be fit with a normal distribution with $\mu_0 = 13.0$ and $\sigma_0^2 = 8.8$. By entering these values along with the sample statistics into the formulas in Table 5.1, the posterior $\theta \sim N(14.78, 0.17)$ is obtained. The prior and posterior are shown in Figure 5.1 and are identified by the number 2 in brackets. It can be seen that within the scale of the figure, the posterior distribution is indistinguishable from that obtained with the uniform noninformative prior. Thus, even though the prior information was used, it was considerably less precise than the sample and had little effect on the posterior.

Suppose that another individual believes that recovery will take somewhat longer and these beliefs are fit with the prior $\theta \sim N(17.00, 11.56)$. The posterior obtained using this prior is $\theta \sim N(14.85, 0.17)$. The prior and the posterior are identified in Figure 5.1 by a number 3 in brackets. Again the posterior is virtually the same as that obtained with the noninformative prior, and even though the priors numbered 2 and 3 are very different, the posteriors are almost the same. Thus the data were sufficiently precise in relation to the priors to bring about a high degree of posterior agreement.

The prior identified by the number 4 in brackets is not imprecise in relation to the sample. This distribution is $\theta \sim N(13.00, 0.35)$, and when combined with the sample statistics, yields the posterior $\theta \sim N(14.21, .12)$. This posterior is different from that obtained with the noninformative prior and the relatively imprecise priors. Whether or not these posteriors would lead to different conclusions or decisions would depend upon the question being asked. If, for example, the main concern was whether or not θ was greater than 20 days, these posteriors would lead to the same conclusion that it is very unlikely that $\theta > 20$. On the other hand, they could lead to different answers for other questions.

COMPARISON OF TWO NORMAL MEANS
WITH KNOWN VARIANCES

Now that the basic ideas of Bayesian inference have been outlined, we will consider how the posterior distribution can be used. In this section we will consider a study involving comparison of a treatment group and a control group. The ideas of the previous chapter are extended to inference for the difference between two normal means, still assuming the variances to be known. Again, this permits us to make a number of points with a minimum of notational and distributional complexity. Subsequent chapters will discuss procedures that can be used when the variance is unknown.

In order to highlight some important features and advantages of the Bayesian approach, the discussion will involve comparison of classical and Bayesian analyses. Features of the Bayesian analysis that have relevance to the utilization of evaluation findings will be emphasized.

ANALYSIS WITH NONINFORMATIVE PRIORS

The Posterior Distribution of the Difference Between Two Means

The analysis is a straightforward extension of the results discussed above. Suppose that we wish to compare a control group (Group 1) and a treatment group (Group 2) in terms of the mean value of some outcome measure in each group. We assume that the observations on the outcome measure are independently obtained from two normal processes or populations with unknown means θ_1 and θ_2, and known variances σ_1^2 and σ_2^2. We will be interested in inference for $\delta = \theta_2 - \theta_1$. The difference between the two means is designated by δ rather than Δ, as in Chapter 2, to emphasize that the difference is no longer considered to be fixed.

Suppose also that our prior information about θ_1 and θ_2 is vague, and that we can approximate our prior information by uniform distributions on θ_1 and θ_2. We will assume that these priors are independent—our information about θ_1 is not influenced by our information about θ_2, and vice versa. Now, if we obtain independent samples of n_1 observations with mean x_1 for Group 1 and n_2 observations with mean $\bar{x}_2$ for Group 2, the posterior distributions of θ_1 and θ_2 will be $\theta_1 \sim N(\bar{x}_1, n_1^{-1}\sigma_1^2)$ and $\theta_2 \sim N(\bar{x}_2, n_2^{-1}\sigma_2^2)$.

Because the priors and the samples are independent, the posteriors are independent also. It can be shown that the difference between two independent normally distributed random variables is also a normally distributed random variable with a mean equal to the difference between the means of the two normal variables, and a variance equal to the sum

TABLE 5.2
Inference for Two Normal Means with
Noninformative Prior Distributions
($\sigma_1{}^2$ and $\sigma_2{}^2$ known)

Prior Distributions:
$\theta_1 \sim$ uniform
$\theta_2 \sim$ uniform
(θ_1, θ_2 independent)

Sample Sufficient Statistics:
$n_1, \bar{x}_1$
$n_2, \bar{x}_2$
(samples independent)

Posterior Distributions:
$\theta_1 \sim N(\bar{x}_1, n_1{}^{-1} \sigma_1{}^2)$
$\theta_2 \sim N(\bar{x}_2, n_2{}^{-1} \sigma_2{}^2)$
$\delta \sim N(\bar{x}_2 - \bar{x}_1, n_1{}^{-1} \sigma_1{}^2 + n_2{}^{-1} \sigma_2{}^2)$

of the variances of the two normal variables. Therefore, the posterior difference between θ_2 and θ_1 is $\delta \sim N(\bar{x}_2 - \bar{x}_1, n_1^{-1}\sigma_1^2 + n_2^{-1}\sigma_2^2)$. These results are summarized in Table 5.2.

Note that two prior distributions are being revised separately here, and then the posteriors are combined to yield the quantity of interest. Alternatively, we could have considered a single joint prior $p(\theta_1, \theta_2)$, multiplied this by the likelihood for the two parameters given the combined samples, obtained a joint posterior, and then derived the distribution of δ from this joint posterior. For present purposes, this would make the analysis more complicated with little substantive advantage. We will talk about prior and posterior belief about δ, but to avoid confusion, keep in mind that the path from prior to posterior involves two separate analyses.

* * * * * *

EXAMPLE 5-1. AN ANALYSIS USING NONINFORMATIVE PRIORS: EVALUATING A MATERNAL-INFANT HEALTH CARE PROGRAM

A health care program is designed to address the problem of low infant birth weight, which is associated with increased infant mortality and morbidity. Pregnant women at risk being served in a hospital outpatient clinic are identified through a screening program using socioeconomic characteristics and behavioral variables such as smoking and alcohol use. Those women agreeing to participate in the study are randomly assigned to either a control condition involving standard prenatal care or to an experimental condition involving nutritional supplement plus education regarding behavioral factors affecting birth weight.[2] Birth weight (measured in grams) is considered to be normally distributed in both groups with unknown mean θ_1 and known variance $\sigma_1^2 = (540)^2$ grams in the control group, and unknown mean θ_2 and known variance $\sigma_2^2 = (500)^2$ grams in the experimental group. Again, we are simplifying the problem here by considering the variances to be known. Birth weight data are obtained for $n_1 = 47$ live births in the control condition and $n_2 = 43$ live births in the experimental condition. The mean birth weight in the control sample is $\bar{x}_1 = 2935$ grams, and the mean birth weight in the experimental group is $\bar{x}_2 = 3065$ grams.

An initial analysis is carried out using noninformative priors. Using the formulas in Table 5.1, the posterior distribution of θ_1, for the control group, is $\theta_1 \sim N[2935, (540)^2/47]$ and the posterior distribution of θ_2, for the experimental group, is $\theta_2 \sim N[3065, (500)^2/43]$. The posterior distribution for δ is

$$\delta \sim N\left[3065 - 2935, \ \frac{(540)^2}{47} + \frac{(500)^2}{43}\right]$$

which reduces to $\delta \sim N[130.00, (109.63)^2]$.

This says that the difference $\delta = \theta_2 - \theta_1$ has a normal distribution with a mean of 130.00 grams and a standard deviation of 109.63 grams. This is the basic inferential result of the study. It is a distribution of belief about the difference of the means in the treatment and control groups. It describes how vague prior belief should be revised on the basis of the data that were obtained. For purposes of summarization and communication, various features of the posterior distribution could be reported. This is considered in the following discussion.

* * * * * *

Inferences from the Posterior Distribution

The features of the posterior that the investigator might choose to report depends upon the question that is being asked. The posterior can be used to answer a variety of questions about the unknown parameter δ. The methods used for answering certain questions are similar, in some respects, to those used in classical testing and estimation. In the Bayesian approach, however, the posterior distribution is the primary result, and testing and estimation for purposes of inference are simply ways of emphasizing certain features of the posterior to answer certain questions. Consequently, these topics have a more subsidiary role in Bayesian inference than they do in classical inference. We will now consider some questions that might be asked in the above study and how they would be addressed in Bayesian and classical terms.

Does the mean of the treatment group exceed that of the control group? This question is addressed by determining the probability that δ exceeds zero. Given that the posterior distribution of δ is normal, the distribution of

$$z = \frac{\delta - (\bar{x}_2 - \bar{x}_1)}{(n_1^{-1} \sigma_1^2 + n_2^{-1} \sigma_2^2)^{\frac{1}{2}}} \qquad [5.10]$$

is standard normal and we can easily make probability statements about whether or not δ exceeds any particular value. The probability that δ exceeds zero is equal to the probability that a standard normal deviate exceeds the value $(0.00 - 130.00)/109.63 = -1.19$. Referring to a cumulative distribution for a standard normal deviate, as tabulated in most statistics texts, we see that this probability is .88. The subjective probability that δ is zero or less is therefore $1.00 - .88 = .12$. Given these probabilities, the odds in favor of δ being greater than zero are $.88/.12$ or 7.33 to 1. Answering such questions given this normal posterior is relatively simple.

In the classical approach this question would be addressed using a one-sided hypotheses test, with the hypotheses H_0: $\delta \leq 0$ and H_1: $\delta > 0$. (To avoid switching back and forth between the Bayesian symbol for the difference δ and the classical symbol Δ, we will use δ throughout this discussion. It is important to keep in mind that in the Bayesian analysis, this difference is random, whereas in the classical analysis it is fixed.) Using the test statistic for the difference between two normal means with known variances, as discussed in Chapter 2, we find $z = 1.19$. The probability that z would exceed this value, if H_0 were true, is .12. Using conventional significance levels of .10, .05, or even .01, this result would not be significant.

There are obvious numerical similarities between the classical and the Bayesian analyses. The p-value of .12, for example, coincides with the posterior probability that H_0 is true. The reason for this is that, in the Bayesian approach, the posterior distribution of δ is such that, for fixed $(\bar{x}_2 - \bar{x}_1)$, the statistic 5.10 has a standard normal distribution, whereas in the classical approach, the sampling distribution of $\bar{x}_2 - \bar{x}_1$ is such that, for fixed $\delta = \delta_0$,

$$z = \frac{(\bar{x}_2 - \bar{x}_1) - \delta}{(n_1^{-1}\sigma_1^2 + n_2^{-1}\sigma_2^2)^{\frac{1}{2}}} \qquad [5.11]$$

has a standard normal distribution. The order of δ and $(\bar{x}_2 - \bar{x}_1)$ is simply reversed in the two different expressions and the z-values have opposite signs. There will be numerical agreement between Bayesian analyses with noninformative priors and classical analyses for many normal theory problems and this provides a convenient basis for comparison. This should, however, be seen as a coincidence rather than as a general feature of these two approaches to inference. For models other than the normal model, this numerical agreement will not necessarily hold. Furthermore, if an informative prior is used in normal theory problems, Bayesian and classical results may not agree numerically.

Even with such numerical agreement, the interpretation is very different. In equation 5.10, it is the unknown parameter δ that is random, and the distribution expresses degree of belief about the size of the difference conditional upon the data. In equation 5.11, on the other hand, it is the statistic $(\bar{x}_2 - \bar{x}_1)$ that is random, and the distribution expresses the relative frequency with which values of a sample statistic would occur in a large number of repeated experiments, conditional upon the difference being equal to some specified value δ_0.

A major advantage of the Bayesian approach is that the probabilities obtained in response to the question pertain directly to what one should believe about whether or not δ exceeds zero. In the classical approach, the observed p-value pertains to the size of the sample statistic that would be obtained over repeated sampling, assuming H_0 to be true. A cutoff value, such as .05, for determining statistical significance simply establishes a limit on the relative frequency of errors that would be made over repeated experiments. As was discussed earlier, this is something less than what users of evaluation findings want to know. P-values and significance levels tell users what they could expect, conducting such a test over identical repeated experiments, but most users want to know what to believe or do given the data at hand. As was discussed in Chapter 2, the classical error probabilities relate to initial precision. The Bayesian posterior probabilities, on the other hand, relate to final precision—what we know after we have carried out the procedure. In routine inspection sampling, for example, such long-run probabilities are directly relevant to the user's problem because the experiment is repeated over and over. In the case of a single study, the relevance of these frequency probabilities for what the user should believe about the hypotheses under consideration is not clear. The Bayesian results, however, apply directly to the question of interest.

Users of classical results sometimes misinterpret classical p-values, in a Bayesian manner, as probability statements about the truth or falsity of hypotheses. Assuming noninformative priors, these users are correct, but for the wrong reason. The classical analysis provides no basis for making probability statements expressing the degree to which a particular hypothesis is true; however, because of the numerical agreement in normal theory problems, these misinterpretations of classical results do coincide with Bayesian conclusions.

A second advantage of the Bayesian analysis can be seen by taking a closer look at the notion of testing. Although the Bayesian approach to the question of whether of not $\delta > 0$ can be thought of in terms of hypothesis testing, Edwards et al. (1963: 214) write, "From a Bayesian point of view, the special role of testing tends to evaporate." They point out that, although hypotheses can be examined within the Bayesian framework, testing of two mutually exclusive hypotheses such as those considered in the classical test would be done by computing the posterior probabilities of the hypotheses; "that a pair of probabilities are singled out for special attention is without theoretical interest" (p. 214). These probabilities are simply features of the posterior distribution, and deriving them does not require any special theory of hypothesis testing.

Actually, examining the relative size of the posterior probabilities for the two hypotheses is not a test in the same sense that the classical hypothesis test is. The classical test is a decision procedure for choosing between the hypotheses in light of the seriousness or costs of error implied in the choice of the decision rule. The Bayesian posterior probabilities for the hypotheses are purely inferential statements and any decision about which of the hypotheses to accept would require some consideration of the cost of error. The Bayesian procedure of reporting the posterior probabilities or odds ratios for hypotheses is more accurately described as a procedure for comparing, rather than testing, hypotheses because no decision to accept or reject is involved (Zellner, 1971: 291-292). Bayesian inference can be used in decision making if desired and this will be considered in a later chapter.

The decision-theoretic nature of the classical test is often overlooked. Yet, as Bakan (1970: 249-250) writes, the contribution of Neyman and Pearson, and of Wald, has been to demonstrate that the methods of testing "constitute not an assertion, or an induction, or a conclusion calculus, but a decision—or risk—evaluation calculus." As soon as one begins to talk about making this choice in such a way as to be appropriately protected against two types of error that differ in seriousness, it is clear that one is attempting to solve a specific decision problem and that the results will be meaningful only in terms of the context of that problem.

For purely inferential problems, the value of a decision procedure is questionable. If the user does not have a particular decision problem at hand, but simply wishes to take the results of the evaluation into account in his or her thinking about an issue or conceptualization of a problem, a decision based on notions of loss is inappropriate. What the user needs to know is how the data modify, or support, his or her beliefs. As Rozeboom (1960) points out, modification of belief is not an all-or-none affair resulting from a decision based on the data. Beliefs are cognitive states that may provide the basis for decisions but are not arrived at by decisions. Instead, the data should confer upon the conclusions "a certain *appropriate degree of belief*" (Rozeboom, 1960: 421). This is precisely what a posterior distribution expresses. As Meyer (1975: 8) points out the Bayesian can provide a full description of the results by reporting the posterior distribution without having to accept or reject anything. Consideration of the posterior draws attention to the magnitude of the treatment effect instead of to a yes-no answer concerning statistical significance. On the other hand, if the user has a true decision problem in which the choice of actions depends upon accepting or rejecting the hypotheses being considered, then only by coincidence will the loss functions implied by arbitrary significance levels be those of the user. Thus statistical significance may not have much practical significance. The classical test as a decision-making procedure will be examined in more detail in Chapter 9. Our point here is that, for purposes of inference, the Bayesian approach provides a much more direct answer to the user's question.

A third advantage to the Bayesian approach stems from the way that the hypotheses are regarded. In the classical test, the null and alternative hypotheses have very different status. Errors in rejecting the null hypothesis when it is true are considered to be much more serious than errors in rejecting the alternative hypothesis when it is true, and in order to ensure that the probability of Type I error is very small, we must be willing to accept a very high probability of Type II error over certain regions of the alternative hypothesis. As we discussed in Chapter 2, this leads to reluctance to accept the null hypothesis when it cannot be rejected. Consequently, failure to reject the null hypothesis often leaves the user in the position of not being able to say much of anything about the unknown parameter. Although this may be acceptable in some basic research where action can be postponed until more data are obtained, this is likely to cause problems in the action setting where action may be necessary on the basis of the data on hand. As was mentioned earlier, in certain evaluation studies, the user might want to conclude that a new, less costly treatment is as good as, or no worse than, a conventional, more costly treatment. Because this involves accepting the null hypothesis, it is is difficult to justify this conclusion. This problem does not arise in the Bayesian approach—the two hypotheses are treated symmetrically. The user can easily obtain a posterior probability for either hypothesis and use this in thinking about any action he or she might want to take.

Does the difference between the treatments exceed some minimum value of practical significance? The question commonly asked in many hypothesis tests on evaluation data is whether or not $\delta > 0$. Indeed, many computer programs only test the null hypothesis $\delta \leq 0$. Yet the answer to this question may not be relevant to the needs of users who actually have decision problems. The relevant question to a user facing a decision may be whether θ_2 exceeds θ_1 by some nonzero amount δ_0, that has some practical significance. Unfortunately, as Roberts (1976: 754) notes, many users of hypothesis tests fail to recognize that the conventional null value and δ_0 need not coincide.

Using the Bayesian posterior distribution it is a simple matter to determine whether or not δ exceeds any particular δ_0 of interest. This is done in the same way as determining whether or not δ exceeds zero. Suppose that the user of these findings felt that δ must exceed 50.00 grams for the treatment to be of much value in his or her setting. The probability is that $\delta > 50.00$ is equal to the probability that a standard normal deviate exceeds $(50.00 - 130.00)/109.63 = -.73$. This probability is .77. Thus the probability that the difference is greater than 50 grams is .77, and the probability that it is not is $1.00 - .77 = .23$. The odds are therefore 3.35 to one in favor of $\delta > 50.00$. In other words, it is about three times as likely that $\delta > 50.00$ than it is that $\delta \leq 50.00$.

A classical one-sided test of the following hypotheses could be carried out: H_0: $\delta \leq 50.00$ versus H_1: $\delta > 50.00$. Here $p < .23$, and this would be judged nonsignificant at conventional significance levels. In practice, this test is seldom carried out. The determination of δ_0 requires some analysis of the user's problem and such analysis is seldom done. Consequently, it is easier to address routinely the question of whether or not $\delta > 0$, than the question of whether or not $\delta > \delta_0$. Unfortunately, the user does not always recognize that the answer that is being provided by the test is for a question other than the one of interest.

It is sometimes argued that testing with a conventional significance level, such as .05, is appropriate because there will be little power to detect small, unimportant differences, and H_0 is not likely to be rejected unless the true difference is truly sizable. It is true that this affords some protection against concluding that $\delta > 0$, when δ is very small. However, as Schlaifer (1959: 655) points out, with conventional significance levels, we will almost always have too much or too little protection. Furthermore, suppose that we do reject H_0. Can we then conclude that $\delta > \delta_0$ with only a .05 chance of error? The answer is no; the significance level pertains to H_0: $\delta < 0$ and says nothing about the probability of error when the true value of δ is greater than zero but less than δ_0. This is because the test is designed to answer the question of whether or not $\delta < 0$, and not the question of whether or not $\delta < \delta_0$. The logic required to justify giving half an answer to the right question by asking the wrong question seems convoluted, to say the least. Questions of this type can be dealt with in a much more straightforward manner in the Bayesian approach.

Note that neither H_0: $\delta \leq 0$ nor H_0: $\delta \leq \delta_0$ could be rejected at any conventional level using the classical test, and it would be hard to justify wider implementation of the program given these results. Yet if the decison maker has given careful consideration to the costs and benefits of implementation versus nonimplementation in arriving at δ_0, statistical significance at any conventional level is irrelevant, and misleading if used as a guide to action. The fact that the odds favor $\delta > \delta_0$ is sufficient for the decision maker's choice. This type of problem is discussed in detail in Chapter 9.

Is there any difference between the treatments? Instead of asking whether θ_2 exceeds θ_1, it is sometimes asked simply: whether they differ. The correct statistical formulation of this question requires some care. If we think about it for a minute, the probability that the difference between the two means is *exactly* zero, is virtually zero. From our posterior distribution of δ, $P(\delta = 0)$ will be zero because the probability of any single point on this continuous distribution is zero. From the Bayesian viewpoint, this is not a very useful question to ask. A more interesting question involves specifying some interval around zero and asking what the probability is that δ lies in that interval. The size of this interval would depend on the magnitude of the difference required for practical significance.

This question figures prominently in classical two-sided tests involving the hypotheses H_0: $\delta = 0$ and H_1: $\delta \neq 0$. We are seldom interested in whether or not δ is exactly equal to zero, and most Bayesians would see this as an inappropriate formulation of the problem. Because we know, to begin with, that δ is not going to be exactly zero, the justification for asking this question is unclear. There are arguments that this test is of value because it tells us something other than what it seems to be designed to tell us; but, as Berger (1980: 22) comments in discussing this test, "It seems somewhat nonsensical, however, to deliberately formulate a problem wrong, and then in an ad hoc fashion explain the final results in more reasonable terms."

There are Bayesian methods of considering the probability that $\delta = 0$ for problems in which there is some concentration of prior probability on zero. These methods will be considered later in this chapter.

In what region does the difference lie? It is sometimes useful for inferential purposes to provide some interval (a, b) in which δ is likely to lie. Again, this is easily obtained from the posterior distribution. For example, suppose that we want to pick the shortest interval in which δ has a 50% chance of lying. Because the normal distribution of δ is symmetrical, the shortest interval will be that including 25% of the area above the mean of 130 grams, and 25% of the area below. From a table of cumulative probabilities for a standard normal distribution, the standardized values that form the upper and lower boundaries of this region are +.67 and −.67, respectively. Thus we have

$$a = 130.00 - (.67)(109.63) = 56.55$$

$$b = 130.00 + (.67)(109.63) = 203.45$$

The subjective probability that δ lies in the interval from 56.55 to 203.45 is .50.

This interval is a 50% highest density region (HDR) of the posterior distribution. The density at any point in this region is greater than any point outside of it. HDRs are also the shortest intervals in which δ is likely to occur with a specified probability. For symmetrical distributions, such as the normal distribution, this will be a central interval and can be simply determined by finding values that contain one-half of the desired percentage of the area on either side of the mean. For asymmetric distributions, special tables may be required, such as those found in Novick and Jackson (1974) and Phillips (1973).

There is some similarity to classical confidence intervals. A 50% confidence interval for the difference δ runs from 56.55 to 203.45. This is identical to the Bayesian 50% HDR. Again, this is due to the similarity between the posterior and the sampling distribution. However, the interpretation is different. In both cases we are saying $P(a \leq \delta \leq b) = .50$ but, in the Bayesian interval, a and b are considered to be fixed and δ is random. The probability is the subjective probability that δ falls into that interval. In the classical interval, the unknown parameter is fixed, and it is the interval (a, b) that is random. The probability is the proportion of times an interval generated by this procedure would contain δ over repeated experiments. Because of this difference in meaning, the Bayesian intervals are called *credible intervals* to distinguish them from confidence intervals. The interpretation of confidence intervals is somewhat counterintuitive, and they are frequently misinterpreted by users as Bayesian credible intervals. Again, from the Bayesian perspective, they are correct, but for the wrong reason. Of course, when informative priors are used, such numerical agreement may not occur. In any case, the Bayesian interpretation is much more direct.

Note also that the endpoints in the classical intervals are dictated by the sample. In the Bayesian approach, we can select any endpoints of interest and determine the probability that δ lies in that interval. For example, the probability that δ will lie in the interval from 50 to 100, say, is $P(\delta > 100) - P(\delta < 50)$, which is $.39 - .23 = .16$.

What is the value of the unknown parameter? If we want a single number that simply gives some idea about the magnitude of the difference, we could select some measure of location or central tendency such as the mean, median, or mode. In the normal distribution, these three measures coincide. The mode of the posterior distribution has the appealing property of having the largest posterior density. However, the choice of an estimator depends upon what you want to do with it. Without such specification, choice of any single number will be somewhat arbitrary. From the Bayesian point of view, the posterior distribution is the full inferential statement about δ, and any single number is inadequate as a description of belief about δ. Any measures of central tendency would need to be accompanied by other information, such as credible intervals or the variance of the posterior, to give an adequate picture of the value of δ. These "estimates," such as the mean, median, and mode, may coincide with the classical estimates arrived at on other grounds, such as unbiasedness. However, classical estimators are selected on the basis of their performance over repeated experiments. The Bayesian estimators are simply descriptive features of the posterior distribution obtained from the data at hand.

If the cost of error is specified, then estimation becomes a decision problem with each action being the choice of a particular value. In the Bayesian approach, we choose as an estimate that value that minimizes expected loss, or maximizes expected utility, with respect to the posterior distribution. With quadratic loss functions, the value that minimizes expected loss is the mean. With linear loss, it is the median. Again, these estimates may coincide with classical estimates obtained through consideration of performance over repeated experiments. However, the Bayesian estimates are tailored to the particular decision problem at hand; they are based on one simple principle: "minimize expected loss" (Zellner, 1971b: 180). It follows directly from the specification of the problem in terms of subjective probability and loss. Classical estimation attempts to provide "good" estimators without specifying how they are to be used. As a result, there are a number of different and sometimes conflicting criteria for what constitutes a good estimator.

Of all the classical estimation procedures, maximum likelihood estimation comes closest in spirit to Bayesian procedures. We see that with a uniform prior, the posterior distribution is proportional to the likelihood function, and the maximum likelihood estimate and the mode of the posterior distribution will be the same. Both approaches share the idea that the import of the data is expressed in the likelihood function. Again, the interpretation is different: The likelihood is the probability of obtaining the data for different values of the parameter, whereas the posterior distribution expresses the probability of different parameter values, conditional upon the data. It is easy to slip often into thinking about the likelihood function as a probability distribution for the parameter, but it is not. The posterior distribution more directly answers the user's questions about the unknown parameter.

ANALYSIS WITH INFORMATIVE PRIORS

The Posterior Distribution of the Difference Between Two Means

In Table 5.1, we summarized how an informative prior of the form $\theta \sim N(\overline{w}, m^{-1}\sigma^2)$ was revised on the basis of a sample with statistics n and $\bar{x}$. Suppose that, in considering two normal means θ_1 and θ_2 with known variances σ_1^2 and σ_2^2 respectively, we have independent prior distributions $\theta_1 \sim N(\bar{w}_1, m_1^{-1}\sigma_1^2)$ and $\theta_2 \sim N(\bar{w}_2, m_2^{-1}\sigma_2^2)$. Then, given the statistics n_1, $\bar{x}_1$, n_2, and $\bar{x}_2$, from two independent samples, we can obtain posterior distributions for θ_1 and θ_2. Because the priors and the samples are independent, the posterior distributions are independent and we can obtain the posterior distribution of δ as discussed in the first section of this chapter. The results are summarized in Table 5.3. While the formulas for the parameters of the posterior distribution of δ are more complex than those in Table 5.2, we can see that the mean of δ is simply the difference between the posterior means of θ_1 and θ_2, and that the variance of δ is the sum of the posterior variances of θ_1 and θ_2.

In the example that we have been considering, there may well be prior information available on birth weight in the form of archival data. We will consider how this information could be used.

*   *   *   *   *   *

EXAMPLE 5-2. AN ANALYSIS USING INFORMATIVE PRIORS: EVALUATING A MATERNAL-INFANT HEALTH CARE PROGRAM

Suppose that a review of medical records over the last year identified 38 women at risk with an average infant birth weight of 2948 grams. This information could be used to construct a prior distribution for the control group, provided that the population served in the clinic has not changed during the past year, and that the standard methods of prenatal care have not changed in that time. If the investigator is satisfied that these conditions hold, analyzing these data using a uniform prior yields a posterior distribution $\theta_1 \sim N[2948, (540)^2/38]$, which can be used, in turn, as a prior distribution for revision with the larger data set discussed above. Entering $m_1 = 38$ and $\bar{w}_1 = 2948$ into formulas for the hyperparameters of the posterior distribution of θ_1, the investigator obtains

$$\theta_1 \sim N\left[\frac{38(2948) + 47(2935)}{38 + 47}, \frac{1}{(38 + 47)}(540)^2\right]$$

which reduces to $\theta_1 \sim N[2940.81, (58.57)^2]$. Suppose also that a small trial of the experimental condition was conducted a few months earlier with a group of women at risk and that birth weights are available

TABLE 5.3
Inference for Two Normal Means with
Informative Prior Distributions
$(\sigma_1^2$ and σ_2^2 known)

Prior Distributions:

$\theta_1 \sim N(\bar{w}_1, m_1^{-1}\sigma_1^2)$

$\theta_2 \sim N(\bar{w}_2, m_2^{-1}\sigma_2^2)$

$(\theta_1, \theta_2$ independent)

Sample Sufficient Statistics:

$n_1, \bar{x}_1$

$n_2, \bar{x}_2$

(samples independent)

Posterior Distributions:

$$\theta_1 \sim N\left[\frac{m_1\bar{w}_1 + n_1\bar{x}_1}{m_1 + n_1}, (m_1+n_1)^{-1}\sigma_1^2\right]$$

$$\theta_2 \sim N\left[\frac{m_2\bar{w}_2 + n_2\bar{x}_2}{m_2 + n_2}, (m_2+n_2)^{-1}\sigma_2^2\right]$$

$$\delta \sim N\left[\frac{m_2\bar{w}_2 + n_2\bar{x}_2}{m_2 + n_2} - \frac{m_1\bar{w}_1 + n_1\bar{x}_1}{m_1 + n_1}, (m_1+n_1)^{-1}\sigma_1^2 + (m_2+n_2)^{-1}\sigma_2^2\right]$$

for 11 live births with a mean of 3037 grams. If the investigator judges these 11 cases to be comparable to those considered in the larger experiment, and the experimental treatment remains the same, these data can be used in constructing a prior distribution for the experimental group. Analyzing these data with a noninformative prior the investigator obtains $\theta_2 \sim N[3037, (500)^2/11]$. Then entering $m_2 = 11$ and $w_2 = 3037$ into formulas for the parameters of the posterior distribution of θ_2 the investigator obtains

$$\theta_2 \sim N\left[\frac{11(3037) + 43(3065)}{(11+43)}, \frac{1}{(11+43)}(500)^2\right]$$

which reduces to $\theta_2 \sim N[3059.30, (68.04)^2]$. Note that the posterior mean is much closer to the sample mean than to the prior mean. This is because the number of sample observations is larger than the number of prior observations.

Because both the priors and sample are independent, the difference $\delta = \theta_2 - \theta_1$ has a posterior distribution that is normal with a mean equal to the difference between the posterior means of θ_2 and θ_1, and a variance equal to the sum of the posterior variances. Thus the posterior distribution of the difference between the treatment and control means is $\delta \sim N[(3059.30 - 2940.81), (58.57)^2 + (68.04)^2]$ or $\delta \sim N[118.49, (89.78)^2]$.

In comparing the posterior here with the posterior obtained with the noninformative priors, we see that the prior information did have some effect on the posterior. In the analysis using the noninformative prior, the mean of δ is 130.00 grams; here it is 118.49 grams. Pooling the prior and sample information here resulted in a somewhat reduced "estimate" of the treatment effect. Note also that the posterior variance of δ obtained using a noninformative prior is $(109.63)^2$. Using an informative prior, it is now $(89.78)^2$. Combining the two sources of information sharpened belief about the value of δ. One consequence of the variance being known that may not be immediately obvious, is that the addition of sample data will always lead to more precise belief. When we consider more realistic models where the variance is unknown, we will see where sample results that are highly discrepant from the prior will tend to reduce posterior precision. This is in accord with common sense— results that strongly diverge from what we believed to begin with should increase our uncertainty.

* * * * * *

Inferences from the Posterior Distribution

The posterior distribution of δ can be used as we discussed above, to make various inferential statements about δ. In this case, however, the posterior reflects prior and sample information, and the results are somewhat different. The probability that $\delta > 0$ is .91, and the probability that $\delta > 50$ is .78. In the case of the noninformative prior, these values were $P(\delta > 0) = .88$ and $P(\delta > 50) = .77$. Even though the mean value of δ is somewhat less when using the informative prior, the increased precision resulting from the combination of prior and sample information has slightly increased the investigator's confidence that δ exceeds these values. Similarly, the 50% HDR now runs from 58.34 to 178.64; its length is 120.30. This is shorter than the 50% HDR obtained with the noninformative priors, which runs from 56.55 to 203.45, a length of 146.90.

The posterior here differs from the posterior distribution obtained with the noninformative prior, and the inferences made from this posterior will not coincide numerically with classical results. The reason for this is the incorporation of prior information. In the classical approach, this information might be used informally in interpreting the results. How this should be done is not specified; the Bayesian method precisely spells out the implications of this prior information.

It is interesting to note that if the study were being conducted to determine whether or not $\delta > 50$, both analyses lead to similar results. In the case of the noninformative prior, $P(\delta > 50) = .77$ and in the case of the informative prior $P(\delta > 50) = .78$. Thus, even though the priors are different and lead to different posteriors, for the purpose of answering this question, the analysis would appear to be fairly robust with regard to modest variation in the prior. The posterior distributions are not identical, however, and may yield different answers to other questions.

A NOTE ON TESTING
POINT NULL HYPOTHESES

Although the point null hypothesis in the problem that was considered in this chapter is of little interest, there are Bayesian methods for dealing with point null hypotheses when this is appropriate. A simple method that is sometimes used is to determine a $100(1 - \alpha)\%$ HDR, observe whether the null value lies inside or outside of this HDR, and reject H_0 at the $100\alpha\%$ level if the null value lies outside of the HDR. If the prior is noninformative, the results will coincide with those of the classical test with significance level α. This method is not accepted by all Bayesians. It has many of the ambiguities of classical significance tests. Furthermore, it does not provide posterior probabilities of the hypotheses, nor does it have a decision-theoretic justification for acceptance or rejection.

There is an alternative that is more in keeping with the Bayesian methods that we have been discussing, and it provides an interesting constrast with the classical two-sided test. Consider the case of a single normal mean θ, where the process or population variance σ^2 is known. We wish to test the hypotheses $H_0: \theta = \theta_0$ and $H_1: \theta \neq \theta_0$. We assign some prior probability P to H_0; this means that the prior has a concentration of probability at the point θ_0. The prior probability of H_1 is therefore $(1 - P)$ and we take this remaining amount of probability to be uniformly distributed over some range of values surrounding the point θ_0. We will approach this problem using the odds-likelihood ratio form of Bayes' theorem discussed in Chapter 4.

The prior odds in favor of H_0 are $\Omega' = P/(1 - P)$. The likelihood function of θ given a sample with sufficient statistics $\bar{x}$ and n, is

$$\exp\left[\frac{n(\bar{x} - \theta)^2}{-2\sigma^2}\right]$$

The likelihood of H_0 given the data is the single value of this function at θ_0:

$$\exp\left[\frac{n(\bar{x} - \theta_0)^2}{-2\sigma^2}\right]$$

The likelihood of H_1 given the data is more complex because H_1 contains a range of values of θ rather than a single point. Therefore it is necessary to integrate the likelihood function with respect to the prior over the range of values in H_1. This yields, roughly speaking, an average of the likelihood function over H_1, and its value is $(2\pi\sigma^2 n^{-1})^{\frac{1}{2}}$. The likelihood ratio is therefore

$$\text{LR} = \frac{\exp\left[\frac{n(\bar{x} - \theta_0)^2}{-2\sigma^2}\right]}{(2\pi\sigma^2 n^{-1})^{\frac{1}{2}}} \qquad [5.12]$$

Multiplying the prior odds Ω' by the likelihood ratio LR, yields the posterior odds Ω'', as described in Chapter 4. Posterior probabilities for these hypotheses can be obtained from the posterior odds, if desired.

Bayesian results using this procedure will not coincide with classical results, and may lead to conclusions that are diametrically opposed to the classical conclusions. Lindley (1957) discusses this issue, and it has come to be known as Lindley's Paradox. It is based on the work of Jeffreys (1961), and has also been called Jeffreys' Paradox by Lindley. Actually it is not a true paradox but rather a point of difference between classical and Bayesian theory. Suppose that we obtain a sample mean $\bar{x}$

such that it is "just significant" in the classical test at the .05 level, say. This means that

$$z = \frac{\bar{x} - \theta_0}{(n^{-1}\sigma^2)^{\frac{1}{2}}} = 1.96 \qquad [5.13]$$

where 1.96 is the critical value z_α for the two-sided test with significance level $\alpha = .05$, as discussed in Chapter 2. From equation 5.13, we obtain $\bar{x} = \theta_0 + 1.96(n^{-1}\sigma^2)^{\frac{1}{2}}$. Substituting this in the likelihood ratio 5.12, we obtain

$$LR = \frac{\exp\left[\dfrac{(1.96)^2}{-2}\right]}{(2\pi\sigma^2 n^{-1})^{\frac{1}{2}}} = \frac{.146 \, n^{\frac{1}{2}}}{(2\pi\sigma^2)^{\frac{1}{2}}}$$

Now consider the Bayesian procedure for determining the posterior odds:

$$\Omega'' = \frac{.146 \, n^{\frac{1}{2}}}{(2\pi\sigma^2)^{\frac{1}{2}}} \times \Omega'$$

The variance σ^2 is known and Ω' is determined by the prior distribution. With these values fixed, it can be seen that Ω'' will vary with n. That is, the meaning of results that are "just significant" and lead to rejection of H_0 in the classical framework will vary substantially in the Bayesian framework, depending upon the sample size involved. With small samples Ω'' may be less than one, and this would be considered evidence against H_0 in both the Bayesian and classical approaches. However, as n increases, Ω'' will at some point become greater than one, and will be interpreted by a Bayesian as evidence in favor of H_0. In other words, the same results that lead a user of the classical test to reject H_0 will be seen by the Bayesian as supporting H_0.

This result is perhaps not all that surprising; users of classical methods acknowledge that with large samples, they are highly likely to obtain results that are significant and to reject any null hypothesis. It has been suggested that investigators should use more stringent significance levels when dealing with large data sets, and that users should carry out informal discounting when reviewing results obtained with large samples. To the Bayesian, these results and the necessity of the ad hoc recommendations cast further doubt on the value of classical hypothesis tests and statistical significance. We will consider the implications of this further in following chapters.

FURTHER READING

The normal theory problem considered in this chapter is also discussed in Box and Tiao (1973), Hays (1973), Hays and Winkler (1971), Lindley (1965b), Novick and Jackson (1974), and Winkler (1972). Raiffa and Schlaifer (1961) develop the notion of conjugate distributions in their comprehensive discussion of Bayesian distribution theory; the theory of conjugate distributions is also discussed in DeGroot (1970). The introductory section of Bernardo (1979) includes references to different approaches to determining noninformative priors and references to discussions of problems that arise with the uncritical use of improper priors. Robustness in the Bayesian framework is reviewed by Berger (1984).

Methods of probability assessment are discussed in the decision theory references listed at the end of Chapters 2 and 9. Hampton et al. (1973) and Berger (1980) classify various methods for assessing prior distributions. Schlaifer (1969) describes methods of assessing and using nonconjugate priors. Winkler (1983) provides an overview of some issues in this area.

Similar comments on classical methods can be found in Anscombe (1961, 1963), Cornfield (1970), Greenwald (1975), Iverson (1970), Jaynes (1976), Lindley and Phillips (1976), Meyer (1964), Phillips (1973), Pitz (1968, 1982), Pratt (1961, 1965), Schlaifer (1959, 1961), Smith (1965), and Zellner (1971b). Barnett (1982) provides an extensive review and comparison of Bayesian, classical, and other approaches. An interesting discussion of hypothesis testing by Bayesians and non-Bayesians can be found in the set of papers by Neyman (1976), Roberts (1976), Kempthorne (1976), and Pratt (1976) stemming from a panel discussion on the topic. Additional discussion of point hypothesis testing and Lindley's Paradox can be found in Edwards et al. (1963). See also Jeffreys (1961), Phillips (1973), and Zellner (1971a).

NOTES

1. Roberts (1977: 156) quotes Savage as saying, "Do research for your enemies, not your friends," and in the Bayesian framework this means designing one's experiments so as to be able to convince those with opposing views that the conclusions are consistent with what would be arrived at using their priors. Jackson et al. (1980) discuss this type of adversarial analysis in planning experiments.

2. Randomization is not essential within the Bayesian framework. It is used in a number of the examples here because it has the advantages of considerably simplifying specification issues and computations. Furthermore, it is one way of justifying to others a lack of experimenter bias. See Berger (1984), Levi (1982), Lindley (1982b), Rubin (1978), and Suppes (1982) for discussion and further references to this area.

6

Inference for Normal Means and Variances

In this chapter, we move to consideration of statistical models that provide more realistic descriptions of phenomena considered in evaluation research. The previous chapter concerned inference for the normal model with known variance. Although the known variance case provides a useful starting point in the discussion of Bayesian inference, it has limited application in evaluation research because the population variances of variables being studied are seldom known. Here, we will examine inferential procedures based on a model where both the mean and variance are unknown. While the general principles of Bayesian inference remain the same as we move from the known to the unknown variance case, the distribution theory and notation becomes more complex because we must work with joint prior and posterior distributions expressing uncertainty about both the mean and variance of the normal distribution. We will consider inferences for single means and variances, and inferences concerning comparisons of means and variances.

The presentation of the examples in this chapter, and in subsequent chapters, will involve both hand analyses and computerized analyses with display of computer output. The computerized analyses are carried out using the Computer-Assisted Data Analysis (CADA) Monitor; both the 1980 and 1983 versions are used. CADA is an interactive Bayesian software package, which includes routines for data management, exploration, analysis, and decision making. It is conversational in nature and is designed to serve both as a teaching tool and a research tool. Novick et al. (1980b, 1983) provide user documentation for the two versions. A description of the 1980 version is provided in Libby et al. (1981).[1]

THE MODEL AND THE
LIKELIHOOD FUNCTION

We start by examining the probability density function for a sample of n independent observations, $x' = (x_1, x_2, \ldots, x_n)$ from a normal process with mean θ and variance ϕ. The symbol ϕ, rather than σ^2, is used to represent the variance to emphasize that it is no longer considered to be a constant. This probability density function is

$$p(x|\theta, \phi) = (2\pi\phi)^{-n/2} \exp\left[\frac{\sum_{i=1}^{n} (x_i - \bar{x})^2 + n(\bar{x} - \theta)^2}{-2\phi}\right]$$

Once the data are obtained, this provides the form of the likelihood function. Because both θ and ϕ are unknown, we will work with the joint likelihood function $\ell(\theta, \phi|x)$, and this has a different form from the likelihood function discussed in Chapter 5. Here we cannot factor the term

$$\sum_{i=1}^{n} (x_i - \bar{x})^2 / -2\phi$$

out of the exponent and absorb it into the constant of proportionality because the variance is no longer a known constant. Nor can $(2\pi\phi)^{-n/2}$ be absorbed into the constant of proportionality for the same reason. Denoting

$$\sum_{i=1}^{n} (x_i - \bar{x})^2$$

by S^2, the kernel of the likelihood for a normal model with unknown variance is therefore written

$$\ell(\theta, \phi | x) \propto \phi^{-n/2} \exp\left[\frac{S^2 + n(x - \theta)^2}{-2\phi}\right] \qquad [6.1]$$

The observations enter the likelihood through the sufficient statistics n, $\bar{x}$, and S^2.

ANALYSIS WITH A NONINFORMATIVE PRIOR

The Prior Distribution

Consider first the situation where one's prior information about θ and ϕ is vague in relation to the information contained in the sample. We will assume that uncertainty about the mean does not depend upon uncertainty about the variance, and vice versa. Our joint prior will then simply be the product of the two noninformative priors for θ and ϕ. As before, we will use a uniform distribution as a noninformative prior for θ. Vagueness about the variance is expressed by a prior that is uniform on $\ln\phi$ the natural logarithm of the variance, rather than on ϕ itself. This noninformative prior for the variance can be arrived at through a number of different arguments. It turns out that, for a variety of models, noninformative priors for parameters ranging from $-\infty$ to $+\infty$ (such as a normal mean) are uniform on the parameter itself, whereas noninformative priors for parameters ranging from 0 to $+\infty$ (such as normal variance) are uniform on the logarithm of the parameter. This log-uniform prior has the appealing feature that if the prior on $\ln\phi$ is uniform, then the prior on the log of the standard deviation $\phi^{1/2}$ and on the log of the precision ϕ^{-1} will also be uniform. Thus the same results will be obtained whether one works directly with ϕ, $\phi^{1/2}$ or ϕ^{-1}.

The likelihood function 6.1 involves ϕ and not $\ln\phi$, and in order to be able to multiply the prior by the likelihood, we need to obtain $p(\phi)$ from p(lnϕ). The relationship between ϕ and lnϕ is such that when p(lnϕ) is uniform, p(ϕ) is proportional to ϕ^{-1}. The joint noninformative prior is the product of the independent noninformative priors p(θ) and p(ϕ), and is therefore

$$p(\theta, \phi) \propto \phi^{-1} \qquad \qquad [6.2]$$

Again, one needs to keep in mind that these noninformative priors are not chosen to be mathematical descriptions or measures of absolute ignorance. Rather, their use allows us to approximate easily the results that would be obtained with a more precise but more time-consuming specification of prior information that is vague in relation to the information provided by the sample.

The Posterior Distribution

Multiplying the likelihood 6.1 by the prior 6.2 yields the following posterior:

$$p(\theta, \phi | x) \propto \phi^{-(n/2)-1} \exp\left[\frac{S^2 + n(x - \theta)^2}{-2\phi}\right] \qquad [6.3]$$

This is a joint distribution of the mean θ and the variance ϕ. The implications of this posterior are most easily understood in terms of the component conditional and marginal distributions for θ and ϕ separately.

The conditional distribution of the mean. For a given value of ϕ the posterior 6.3 reduces to

$$p(\theta | x, \phi) \propto \exp\left[\frac{n(\bar{x} - \theta)^2}{-2\phi}\right]$$

This is a kernel of a normal distribution with mean $\bar{x}$ with variance $n^{-1}\phi$. That is, conditional upon ϕ, $\theta \sim N(\bar{x}, n^{-1}\phi)$. This is the same as the result we obtained in the known variance case discussed in the previous chapter. We could also obtain the conditional distribution of ϕ for a given value of θ, but it is not needed in this presentation.

The marginal distribution of the variance. Deriving the marginal distribution of ϕ requires integral calculus and we simply present the result. The marginal distribution of the variance is an inverse chi-square distribution. This is a two-parameter distribution; following Novick and Jackson, we write $\phi \sim \chi^{-2}(\nu, \lambda)$. The first parameter is the degrees of freedom associated with the distribution and, in this case, is equal to $n - 1$. The second parameter is a scale factor, which affects the spread of the distribution and in this case is equal to S^2. Thus with noninformative priors, we write the posterior distributions as $\phi \sim \chi^{-2}(n - 1, S^2)$.

The inverse chi-square distribution is really just another way of expressing a relationship that is similar to one that we have already discussed. In Chapter 2 we saw that for the normal model with unknown mean, S^2/σ^2 has a chi-square distribution on $n - 1$ degrees of freedom; this provides the basis for classical inference about σ^2. In Bayesian inference for ϕ with noninformative priors, it also turns out that S^2/ϕ has a chi-square distribution on $n - 1$ degrees of freedom. Here, however, S^2 is fixed and ϕ is the random quantity. Because ϕ is the quantity of interest, it would be easier to talk about a distribution for ϕ rather than one for S^2/ϕ. If the quantity $Y = S^2/\phi$ has a chi-square distribution on $n - 1$ degrees of freedom, then $\phi = S^2/Y$, is said to have an inverse chi-square distribution on $n - 1$ degrees of freedom with scale factor S^2. The inverse chi-square distribution simply enables us to deal more directly with the unknown quantity of interest, ϕ.

We will often describe prior and posterior distributions in terms of the standard deviation $\phi^{1/2}$, rather than the variance ϕ. If $\phi \sim \chi^{-1}(\nu, \lambda)$, then $\phi^{1/2}$ has an inverse chi distribution on ν degrees of freedom with a scale parameter of $\lambda^{1/2}$. Following Novick and Jackson, we write $\phi^{1/2} \sim \chi^{-1}(\nu, \lambda^{1/2})$. In the case of vague prior information, the prior marginal distribution on $\phi^{1/2}$ will be log-uniform. The marginal posterior distribution of the standard deviation is $\phi^{1/2} \sim \chi^{-1}(n - 1, S)$, where S is the square root of the sum of squared deviations S^2.

The marginal distributions for ϕ and $\phi^{1/2}$ enable us to make inferences about these quantities without having to specify a particular value for the mean. These distributions are skewed when the degrees of freedom are small, but become more symmetric and bell-shaped as ν increases.[2]

The reader should be aware that the problem of inference for the unknown variance is discussed in different terms by different writers. Novick and Jackson (1974) and Box and Tiao (1973) work with χ^{-1} and χ^{-2} distributions as described here; other writers work with the related gamma distribution and variations of it. Furthermore, some writers focus on inference for the variance, others on inference for the standard deviation, and some are concerned with inference for the precision, or inverse, of the variance. The reader should be careful to ascertain which of these is being considered.

A joint distribution of two variables can be obtained from the product of the marginal and conditional distributions. The joint distribution 6.3 can be seen as the product of the marginal distribution of the variance and the conditional distribution of the mean, and it will be referred to as a normal inverse chi square distribution.

The marginal distribution of the mean. In making inferences about the mean, we will often want to be able to make statements about the value of θ irrespective of the value of the variance, and the normal conditional distribution of θ given ϕ is not appropriate for this. Instead, we will need the marginal unconditional distribution of the mean. Again, this marginal distribution is obtained using integral calculus. In obtaining this marginal distribution for θ we integrate out the process parameter ϕ so that we can make inferential statements without specifying a particular value for ϕ. The resulting marginal distribution of θ is an unstandardized t distribution. It has three parameters: ν, the degrees of freedom; ζ, the mean; and κ, the scale factor. This is written $\theta \sim t(\nu, \zeta, \kappa)$. In the case of noninformative priors, the parameters of the posterior marginal distribution of θ are: $\nu = n - 1$, $\zeta = \bar{x}$, and $\kappa = n^{-1}S^2$; thus we write $\theta \sim t(n - 1, \bar{x}, n^{-1}S^2)$. The relationship among these parameters is such that

$$t = \frac{\theta - \bar{x}}{[n^{-1}S^2/(n - 1)]^{\frac{1}{2}}}$$

has a standardized t distribution with n – 1 degrees of freedom. Here again the similarity to the classical sampling distribution discussed in Chapter 2 can be seen. The interpretation, though, is quite different: Here $\bar{x}$ is fixed and θ is random.

In classical texts, the t distribution is sometimes described as a one-parameter distribution. Indeed, we only need one parameter, the degrees of freedom, to look up critical values of t in a table after conducting a t test. This is because the t statistic obtained in the test has a standardized distribution, and the tabled values are those of a standardized t distribution with mean equal to zero and $(\kappa/\nu)^{\frac{1}{2}}$ equal to one. In

Bayesian inference we will need to use the unstandardized distribution and explicitly deal with all three parameters. The variance of the t distribution is $\kappa/(\nu - 2)$. When ν is large, the t distribution can be approximated by a normal distribution with mean ζ and variance $\kappa/(\nu - 2)$.

Although these marginal distributions greatly aid in interpreting the posterior distribution, their use requires some care. In this case, they are not independent—the joint distribution is not simply a product of the marginal distributions. This means that we cannot compute posterior probabilities for θ and ϕ from their respective marginal distributions, and then make joint probability statements about θ and ϕ by multiplying the two marginal probabilities. This presentation will focus primarily on one parameter or the other and will make inferences from a single marginal distribution. Readers interested in making joint inferences will find methods described in Box and Tiao (1973).

The inferential process that we have been discussing is summarized in Table 6.1. The table shows the prior marginal distributions, the sample sufficient statistics, and the posterior marginal distributions. We will be primarily concerned with inferences from the marginal distributions and therefore show the prior and posterior marginal distributions. (Keep in mind that in the revision process, it is a joint prior that is being combined with the likelihood to yield a joint posterior. These marginal distributions are secondary features of these joint distributions. The marginal distributions are not being revised independently.) In this case with the noninformative prior, the posterior is determined entirely by the sample, and the hyperparameters of the posterior distribution involve the sample statistics only.

$$* \quad * \quad * \quad * \quad * \quad *$$

EXAMPLE 6-1. AN ANALYSIS USING A NONINFORMATIVE PRIOR: EVALUATING A JOB SKILLS TRAINING PROGRAM

A follow-up study of persons participating in a job skills training program is conducted one year after completing the program. A lengthy interview is carried out with a sample of the participants. In one section of the interview, those persons who are employed are asked to report their gross earnings for the previous week. One aspect of the evaluation focuses on the average gross weekly earnings for trainees. The investi-

TABLE 6.1
Inference for a Mean and Standard Deviation of a Normal
Distribution with a Noninformative Prior Distribution

Prior Marginal Distributions:

$\theta^{\frac{1}{2}} \sim$ log uniform
$\theta \sim$ uniform
$(\theta, \phi^{\frac{1}{2}}$ independent)

Sample Sufficient Statistics:

$n, \bar{x}, S^2$

Posterior Marginal Distributions:

$\phi^{\frac{1}{2}} \sim \chi^{-1} (n - 1, S)$
$\theta \sim t(n - 1, \bar{x}, n^{-1} S^2)$

---

gator considers earnings to be normally distributed and decides to analyze the data with a noninformative prior.

We neeed to mention one point before discussing the analysis. The sample sum of squared deviations S^2 and the square root of this quantity S appear in a number of calculations. However, the sample standard deviation is a more easily interpreted measure of sample dispersion, and in the following examples we will present the sample standard deviation and derive S^2 or S from it. The sample standard deviation is defined here as $s_n = (n^{-1}S^2)^{\frac{1}{2}}$. Note that the divisor is n rather than n − 1. Because unbiased estimation is not a consideration within the Bayesian framework, Bayesians sometimes find it more convenient to work with a divisor of n. To avoid any confusion, we will use s_n to denote the sample standard deviation calculated using n. The CADA programs that we will work with use s_n.

Data are obtained from 32 respondents, with the mean earnings being $255.81, with a standard deviation $s_n = \$17.15$. Because $s_n = 17.5$, then $S^2 = 32(17.5)^2 = 9411.92$. The sample sufficient statistics are, therefore, n = 32, x = 255.81, and $S^2 = 9411.92$.

Given the noninformative prior distribution, and these sample statistics, the formulas in Table 6.1 yield the following marginal posterior distributions for the standard deviation $\phi^{\frac{1}{2}}$ and the mean θ:

$$\phi^{\frac{1}{2}} \sim \chi^{-1}(32 - 1, 9411.92^{\frac{1}{2}})$$

which is $\phi^{\frac{1}{2}} \sim \chi^{-1}(31, 97.02)$; and

$$\theta \sim t(32 - 1, 255.81, (32)^{-1} 9411.92)$$

```
OPTION 6:  GRAPH OF THE DENSITY FUNCTION   OVER 99% HDR
              STUDENT'S T DISTRIBUTION
DEGREES OF FREEDOM =     31.00                MEAN =    255.81
SCALE PARAMETER    =    294.12   STANDARD DEVIATION =      3.18
------------------------------------------------------------
THESE ARE THE PARAMETERS OF THE DISTRIBUTION TO BE GRAPHED.
WHEN YOU ARE READY FOR THE GRAPH TO BE DISPLAYED TYPE '1'? 1

    DF=     31.00   MEAN=      255.81   ST.DEV=       3.18
   247.36 I\
   248.20 I\\
   249.05 I\\\\
   249.89 I\\\\\\\\
   250.74 I\\\\\\\\\\I\\\
   251.58 I\\\\\\\\\\\I\\\\\\\\\\
   252.43 I\\\\\\\\\\\I\\\\\\\\\\\\\I\\\\\\\\
   253.27 I\\\\\\\\\\\I\\\\\\\\\\\\\\I\\\\\\\\\\\\\\\I\\\\\\
   254.12 I\\\\\\\\\\\I\\\\\\\\\\\\\\\I\\\\\\\\\\\\\\\\\I\\\\\\\\\\\\\\\I\\
   254.96 I\\\\\\\\\\\\I\\\\\\\\\\\\\\\\\I\\\\\\\\\\\\\\\\\I\\\\\\\\\\\\\\\I\\\\\\\\\\\\
   255.81 I>>>>>>>>>>1>>>>>>>>>>2>>>>>>>>>>3>>>>>>>>>>4>>>>>>>>>5
   256.66 I////////////I////////////I////////////I////////////I/////////
   257.50 I////////////I///////////I////////////I///////////I//
   258.35 I////////////I///////////I///////////I//////
   259.19 I////////////I//////////I///////
   260.04 I///////////I/////////
   260.88 I///////////I///
   261.73 I/////////
   262.57 I////
   263.42 I//
   264.26 I/
```

SOURCE: Output generated using CADA (Novick et al., 1980b).

Figure 6.1 Posterior Distribution of Mean Weekly Earnings θ

which is $\theta \sim t(31, 255.81, 294.12)$. These results could have been obtained using CADA by either accessing a raw data file or entering the sample statistics.

Consider first inference for the unknown mean θ. The posterior distribution will be evaluated using CADA. Once one enters the degrees of freedom, the mean, and the scale parameter of this unstandardized t distribution, a number of its features can be easily examined. One can choose among various options to obtain: percentiles, highest density regions, probabilities t is above and below some value, probability t is between two values, and a graph of the density function.

The graph of the posterior distribution shown is Figure 6.1. This is the distribution of belief about θ when prior belief is vague. Such a graph can be useful in reporting; one can easily see where posterior belief is concentrated.

Certain features of this distribution might also be of interest. Using CADA, one might request the boundaries of the 95% HDR; the result is the interval (249.53, 262.09). This same result could be obtained by hand. The 97.5 percentile of a standardized t distribution on 31 degrees of freedom is 2.04; 95% of this distribution lies between −2.04 and +2.04. Thus

$$P[-2.04 \leq (\theta - 255.81)/(294.12/31)^{1/2} \leq +2.04] = .95$$

and

$$P(249.53 \leq \theta \leq 262.09) = .95.$$

Because the t distribution is symmetric, this central interval is an HDR. The person with vague prior information should be 95% certain that the mean weekly earnings of the trainees lies between these two values. The odds are 19 to 1 that θ lies in this interval. These boundaries of this 95% HDR are the same as those of the classical 95% confidence interval, but the interpretation is very different. Other HDRs might also be useful in reporting. The 75% HDR is the interval (252.20, 259.42); the 50% HDR is the interval (253.71, 257.91); and so on. The reporting of selected HDRs may aid in the interpretation of the plot of the posterior.

Another feature that could be reported is the probability that θ is greater or less than some value of interest. Suppose that the investigator is interested in the probability that θ exceeds some comparison value of $260.00, say. Using CADA one obtains $P(\theta > 260.00) = .09$ and $P(\theta < 260.00) = .91$. This could also be obtained by hand:

$$P(\theta > 260.00) = P[t > (260 - 255.81)/(294.12/13)^{1/2}] = P(t > 1.36)$$

where t is a standardized t variate with 31 degrees of freedom. The value 1.36 is simply the standardized value of 260.00. The probability that t exceeds 1.36 is .09, and the probability that it is less than 260.00 is $(1.00 - .09) = .91$. It might be useful to report these probabilities for a number of values. Table 6.2 shows the probability that θ is greater or less than a number of different values. A user of these findings whose prior beliefs are vague could simply read off the probabilities for any value of interest. Suppose someone is interested in the probability that θ exceeds 250.00. (Weekly earnings of $250.00 per week translate into annual earnings of $13,000.00.) This probability is .97. The odds are over 32 to 1 that θ exceeds 250.00.

A user of classical methods might address this question using a hypothesis test with the hypotheses $H_0: \theta \leq 250.00$ and $H_1: \theta > 250.00$. Using the test statistic described in Chapter 2, we find $p < .03$. In other words, if H_0 is true, the probability of obtaining a value of 255.81 or larger is less than .03. This coincides numerically with the Bayesian

TABLE 6.2
Probability That Mean Weekly Earnings θ is
Greater Than and Less Than Selected Values c

c	$P(\theta > c)$	$P(\theta < c)$
246	1.00*	.00*
248	.99	.01
250	.97	.03
252	.89	.11
254	.72	.28
256	.48	.52
258	.24	.76
260	.09	.91
262	.03	.97
264	.01	.99
266	.00*	1.00*

*Rounded values.

posterior probability that $\theta < 250.00$; however, these probabilities have different meanings. The classical result is significant at the .05 level. Keep in mind, however, that the classical test is a decision procedure and implicit in the choice of the significance level is some notion of loss. The Bayesian analysis just described simply provides a posterior probability; whether or not $P(\theta > 250.00)$ is "large enough" depends on how the user plans to use the information.

Although the focus has been on inference for the mean, one might also want to consider the marginal posterior distribution for $\phi^{1/2}$. A graph of this inverse-chi distribution is shown in Figure 6.2. This expresses the distribution of belief about the standard deviation after obtaining the data. The variation in earnings among trainees may well be of interest in evaluating the effects of the training program, and the investigator might want to report certain features of this distribution using summary statements similar to those just described.

The inverse chi distribution is asymmetric. Consequently, one cannot obtain $100(1 - \alpha)\%$ HDR's simply by finding values that cut off $100(\frac{1}{2}\alpha)\%$ of the distribution in each tail. The probability of $\phi^{1/2}$ lying in an interval determined in this manner would indeed be $(1 - \alpha)$; however, this credible interval would not necessarily be the shortest interval, as is the HDR. HDRs for this distribution can be easily obtained using CADA. The 50% HDR of the posterior distribution of $\phi^{1/2}$ in Figure 6.2 is the interval (15.76, 18.75), and the 90% HDR is the interval (14.07, 21.53). Novick and Jackson provide special tables for determining HDRs without computer assistance. Classical two-sided confidence intervals are commonly obtained by finding values that cut off equal areas in each tail of the χ^2 distribution, and will therefore not coincide with the Bayesian HDR.

Suppose that the investigator were interested in whether or not $\phi^{1/2} > 20.0$. This probability is equal to the probability that a standardized χ^{-1} variate exceeds $20.00/97.02 = .212$. From CADA, this probability is found to be .21. In the classical framework this question could be addressed by testing the hypothesis

$$H_0: \phi^{1/2} \leq 20.00$$

$$H_1: \phi^{1/2} > 20.00$$

The p-value here is the probability that a chi-square variate with 31 degrees of freedom exceeds $(97.02)^2/(20.00)^2 = 23.53$, which is .79. Again, the results coincide numerically but the meaning is different.

$$* \quad * \quad * \quad * \quad * \quad *$$

ANALYSIS WITH AN INFORMATIVE PRIOR

The natural conjugate family for the normal model with unknown variance is the family of distributions that are jointly normal and inverse chi-square, which is the form of the posterior just discussed. If a prior is a member of this family, then the posterior will be a member also.

A Data-Based Prior

As we discussed in the previous chapter, the posterior obtained in one study can be used as a prior in a subsequent study. Suppose we had obtained a previous sample of m observations with mean $\overline{w}$ and sum of squared deviations R^2. If we analyzed these data with a noninformative prior distribution, the posterior distribution is normal inverse chi squared and has the marginal distributions $\phi^{1/2} \sim \chi^{-1}(m - 1, R)$ and $\theta \sim t(m - 1, \overline{w}, m^{-1}R^2)$, where R is the square root of the sum of squared deviations R^2. If we then use this posterior as a prior distribution in a subsequent study, it is in the very form that is needed for inference using natural conjugate distributions. Note that here we have m observations worth of prior information of both θ and $\phi^{1/2}$. In subsequent problems that will be considered, this will not necessarily be the case.

Suppose we obtain a sample in the subsequent study with sufficient statistics n, $\overline{x}$, and S^2. Inference using the informative prior with these

```
OPTION 5:  GRAPH OF THE DENSITY FUNCTION OVER 99% HDR
-----------------------------------------------------------
               INVERSE-CHI DISTRIBUTION
DEGREES OF FREEDOM =   31.00     SCALE PARAMETER =    97.020
            MEAN =   17.86        STAN. DEV. =     2.352
-----------------------------------------------------------
THESE ARE THE PARAMETERS OF THE DISTRIBUTION TO BE GRAPHED.
WHEN YOU ARE READY FOR THE GRAPH TO BE DISPLAYED TYPE '1'.? 1

   DEGREES OF FREEDOM =    31.00    SCALE =    97.02
   12.73 I\
   13.37 I\\\\
   14.01 I\\\\\\\\\\I\
   14.65 I\\\\\\\\\\\I\\\\\\\\\\I
   15.29 I\\\\\\\\\\I\\\\\\\\\\\\I\\\\\\\\\\\I\
   15.93 I\\\\\\\\\\I\\\\\\\\\\\\\I\\\\\\\\\\\\I\\\\\\\\\\\\I\
   16.57 I\\\\\\\\\\I\\\\\\\\\\\\I\\\\\\\\\\\\I\\\\\\\\\\\\I\\\\\\\\\
   17.20 I>>>>>>>>>1>>>>>>>>>>2>>>>>>>>>>3>>>>>>>>>>4>>>>>>>>>
   17.84 I//////////I//////////I//////////I//////////I///////
   18.48 I//////////I//////////I//////////I//////////I//
   19.12 I//////////I//////////I//////////I/////
   19.76 I//////////I//////////I///////
   20.40 I//////////I//////////I/
   21.04 I//////////I/////
   21.68 I//////////I/
   22.32 I///////
   22.96 I/////
   23.60 I///
   24.24 I//
   24.88 I/
```

SOURCE: Output generated from CADA (Novick et al., 1980).

Figure 6.2 Posterior Distribution of the Standard Deviation of Weekly Earnings $\phi^{1/2}$

data is summarized in Table 6.3. The prior and posterior marginal distributions are shown so that one can see how they change from prior to posterior. Looking at the posterior distributions, we see that the degrees of freedom for θ and $\phi^{1/2}$ are based on the combined number of prior and sample observations. It can also be seen that, as in known variance cases, the posterior mean of θ is a weighted average of the prior and sample means. Finally, the sum of squares terms in the scale parameters of θ and $\phi^{1/2}$ are composed of the combined prior and sample sums of squares, plus a between prior and sample sum of squares. The difference between the prior mean $\overline{w}$ and the sample mean $\overline{x}$ contributes to the size of the scale factor and to the dispersion of the posterior distribution. Differences between the prior and sample means provide some information about ϕ. This makes sense—if the sample differs from the prior, this should contribute to posterior uncertainty. Indeed with sizable differences, the posterior may express increased rather than decreased uncertainty after obtaining the sample.

*   *   *   *   *   *

TABLE 6.3
Inference for a Mean and Standard Deviation of a Normal
Distribution with an Informative Prior Distribution

Prior Marginal Distributions:

$\theta^{1/2} \sim \chi^{-1}\ (m-1, R)$

$\theta \sim t(m-1, \bar{w}, m^{-1}\ R^2)$

Sample Sufficient Statistics:

$n, \bar{x}, S^2$

Posterior Marginal Distributions:

$$\phi^{1/2} \sim \chi^{-1}\left[m+n-1, (R^2+S^2+\frac{mn\,(\bar{x}-\bar{w})^2}{m+n})^{1/2}\right]$$

$$\theta \sim t\left[m+n-1,, \frac{m\bar{w}+n\bar{x}}{m+n},\ (m+n)^{-1}\ (R^2+S^2+\frac{mn\,(\bar{x}-\bar{w})^2}{m+n})\right]$$

EXAMPLE 6-2. AN ANALYSIS USING A DATA-BASED INFORMATIVE PRIOR: EVALUATING A JOB SKILLS TRAINING PROGRAM

To continue the previous example, a second sample is obtained one month later, and is composed of trainees who completed their training a month after the first group. The investigator judges this second sample to be comparable to the first sample, and notes that there have been no major changes in the local job market or economy that would affect the earnings of the second group in a manner different from the first group.

The investigator uses the posterior distribution obtained with the first sample as a prior distribution. The second sample contains n = 27 observations with a mean of $\bar{x}$ = 259.78 and a standard deviation of s_n = 16.03. From the standard deviation, we determine S^2 to be 6937.94. From the posterior of the previous study we have m = 32, $\bar{w}$ = 255.81, and R^2 = 9411.92. Entering these sample and prior quantities into the formulas for the parameters of the posterior distributions one obtains

$$\theta^{1/2} \sim \chi^{-1}\left[32+27-1, \left(9411.92+6937.94+\frac{(32)(27)(259.78-255.81)^2}{32+27}\right)^{1/2}\right]$$

which reduces to $\phi^{1/2}: \chi^{-1}$ (58, 128.77), and

$$\theta \sim t \left[32 + 27 - 1, \ \frac{32(255.81) + 27(259.78)}{32 + 27}, \ (32 + 27)^{-1} (128.77)^2 \right]$$

which reduces to $\theta \sim t$ (58, 257.63, 281.03).

These posterior distributions can also be obtained using CADA. The investigator could enter the hyperparameters of the prior marginal distributions and the sample sufficient statistics, and obtain the summaries shown in Table 6.4. The summaries contain features of both the prior and posterior distributions for comparative purposes. We can see that the posterior HDRs for both θ and $\phi^{1/2}$ are shorter than the prior HDRs. The additional information in the second sample has led to more precise belief about the values of the unknown parameters.

*   *   *   *   *   *

A Non-Data-Based Prior

In the previous example, prior information consisted of the results of the earlier study, and these results were already in natural conjugate form. Specification of a prior distribution was therefore fairly simple. In other problems this specification may be less straightforward. Although an individual's prior beliefs may reflect previous data, they may not be entirely determined by these data. What is needed is some method for eliciting the individual's prior beliefs in a form that can be introduced into the analysis.

We will illustrate one method of doing this using CADA. Although one can construct joint priors without computer assistance (see Novick and Jackson, 1974; Pitz, 1982), this process is greatly facilitated by computerized methods. As we will see, assessment is an iterative process. Most people do not have precisely defined prior probability distributions in their heads. In the assessment process, the individual specifies certain features of his or her prior beliefs, and then examines the implications of these features. The individual is forced to confront any inconsistency or incoherence. In resolving incoherence, the individual may alter some aspects of the original specification and then look

TABLE 6.4
Summaries of Prior and Posterior Distributions of $\phi^{1/2}$ and θ

*** SUMMARY OF ANALYSIS ON THE STANDARD DEVIATION ****

MARGINAL INVERSE CHI DISTRIBUTIONS

	PRIOR			POSTERIOR		
DEGREES OF FREEDOM	31.00			58.00		
SCALE PARAMETER	97.02			128.77		
MEAN	17.86			17.13		
MODE	17.15			16.76		
MEDIAN	17.62			17.01		
50% HDR	15.77	TO	18.75	15.76	TO	17.88
75% HDR	14.90	TO	20.04	15.11	TO	18.74
95% HDR	13.59	TO	22.59	14.10	TO	20.38

TO CONTINUE,   TYPE '1'.? 1

********* SUMMARY OF ANALYSIS ON THE MEAN ************

MARGINAL STUDENT'S DISTRIBUTIONS

	PRIOR			POSTERIOR		
DEGREES OF FREEDOM	31.00			58.00		
SCALE PARAMETER	294.12			281.03		
MEDIAN	255.81			257.63		
50% HDR	253.71	TO	257.91	256.13	TO	259.12
75% HDR	252.20	TO	259.42	255.07	TO	260.18
95% HDR	249.53	TO	262.09	253.22	TO	262.03

SOURCE: Output generated using CADA (Novick et al., 1980b).

at the implications again. Doing this by hand can be laborious; an interactive computer program is ideally suited for this type of activity. In the example that will be considered, CADA will be used to fit a natural conjugate distribution to an individual's beliefs.

In fitting non-data-based priors, it is useful to consider an extension of the natural conjugate family discussed here. In the formulas in Table 6.3, it is assumed that whether we are working with an actual prior sample or with an equivalent prior sample, there is one value m for the prior sample size. The form that the conjugate prior can assume can be expanded by allowing the number of prior observations providing information about $\phi^{1/2}$ to differ from the number of prior observations providing information about θ. The sample sizes associated with the standard deviation and the mean will be denoted by m_S and m_M, respectively. As Novick and Jackson point out, we can think of the prior information as arriving from two prior samples, one of which provides information on $\phi^{1/2}$ only, and the other that subsequently provides some information on both $\phi^{1/2}$ and θ. Combining these two samples using Bayes' theorem yields a distribution of the desired form. This distribution can be used as a prior just as in the case where $m_S = m_M = m$. Allowing m_S and m_M to differ provides for greater flexibility in fitting prior belief with a member of the natural conjugate family. There is a restriction, however, that $m_S \geq m_M$. Additional flexibility can be obtained by allowing m_S and m_M to assume noninteger values. We can use the formulas in Table 6.3 with this extended conjugate distribution provided that we substitute m_S for m in the degrees of freedom parameters and substitute m_M for m elsewhere.

* * * * * *

EXAMPLE 6.3. AN ANALYSIS USING A NON-DATA-BASED INFORMATIVE PRIOR: EVALUATING A JOB SKILLS TRAINING PROGRAM

Consider a modification of the preceding example. Suppose that the requirements for admission into the training program had been modified slightly when the second group entered the program. This second group is somewhat more heterogeneous on a number of demographic characteristics; however, it is not clear what effect this would have on subsequent earnings. The investigator feels that, although the two groups are still generally similar, he is somewhat less certain about the values of θ and $\phi^{1/2}$ than he would be if he felt the two samples were equivalent. The investigator is involved in the development of this training program, and wishes to carry out an analysis involving an assessment of his prior in order to guide his thinking. (If the investigator were conducting the study for a specific client, he might want to assist the client in specifying the client's prior beliefs.)

In assessing the joint prior distribution using CADA, the user is first assisted in specifying an inverse chi distribution for the standard deviation. Following this, the user is assisted in specifying a normal distribution for the mean, conditional upon the median value of the χ^{-1} distribution for $\phi^{1/2}$. The program then generates the marginal t distribution for θ.

The use of CADA to fit prior beliefs about $\phi^{1/2}$ with an inverse chi distribution is illustrated in Exhibit 6.1. The steps are self-explanatory; however, we will note some aspects of the user's thinking as the program

EXHIBIT 6.1

Assessment of a Prior Marginal Distribution for $\phi^{1/2}$

PRIOR DISTRIBUTION ON THE STANDARD DEVIATION

THIS MODULE WILL ASSIST YOU IN FITTING AN INVERSE CHI DISTRIBUTION
TO YOUR PRIOR BELIEFS ABOUT THE STANDARD DEVIATION OF A NORMAL
DISTRIBUTION

WE BEGIN BY ASKING YOU TO SPECIFY THE 25TH, 50TH AND
75TH PERCENTILES OF YOUR PRIOR DISTRIBUTION.

SPECIFY 50TH PERCENTILE.   YOUR BETTING ODDS ARE EVEN THAT THE   **[1]**
STANDARD DEVIATION IS LESS THAN THIS VALUE. INPUT 50TH.? 18

SPECIFY 25TH PERCENTILE.   YOUR BETTING ODDS ARE 3 TO 1 THAT THE   **[2]**
STANDARD DEVIATION IS GREATER THAN THIS VALUE. INPUT 25TH.? 16.5

SPECIFY 75TH PERCENTILE.   YOUR BETTING ODDS ARE 1 TO 3 THAT THE
STANDARD DEVIATION IS GREATER THAN THIS VALUE. INPUT 75TH.? 19.8

FOUR POSSIBLE APPROXIMATE PRIOR DISTRIBUTIONS ARE NOW BEING
COMPUTED FOR YOUR CONSIDERATION.

HERE ARE THE PERCENTILES OF FOUR INVERSE CHI DISTRIBUTIONS
FITTED TO YOUR PERCENTILE SPECIFICATIONS.

	10TH	25TH	50TH	75TH	90TH
1	15.46	16.58	18.00	19.63	21.33
2	15.42	16.59	18.08	19.79	21.58
3	15.24	16.45	18.00	19.81	21.69
4	15.32	16.51	18.04	19.82	21.68

COMPARE THE PERCENTILES OF THESE DISTRIBUTIONS AND DECIDE
WHICH MOST CLOSELY CORRESPONDS TO YOUR PRIOR BELIEFS. YOU
CAN EITHER TENTATIVELY ACCEPT THIS DISTRIBUTION OR SPECIFY
NEW VALUES FOR THE PERCENTILES.

IF YOU WANT ONE OF THE DISTRIBUTION, TYPE ITS NUMBER.   **[3]**
IF YOU WANT TO RESPECIFY THE PERCENTILES,   TYPE '0'.? 4

HERE ARE SOME OF THE CHARACTERISTICS OF THE INVERSE CHI   **[4]**
DISTRIBUTION YOU ARE NOW CONSIDERING.

HYPOTHETICAL SAMPLE SIZE(M)	29.00	
10TH PERCENTILE	15.32	
25TH PERCENTILE	16.51	
50TH PERCENTILE	18.04	
75TH PERCENTILE	19.82	
90TH PERCENTILE	21.68	
50% HDR	16.03 TO	19.25
75% HDR	15.10 TO	20.65
95% HDR	13.72 TO	23.44

EXHIBIT 6.1 Continued

IF YOU DO NOT FEEL THAT THIS HYPOTHETICAL SAMPLE SIZE (M)
REFLECTS YOUR PRIOR INFORMATION ABOUT THE STANDARD DEVIATION
YOU CAN SPECIFY A DIFFERENT ONE. A DIFFERENT M WILL NOT AFFECT
THE MEDIAN, BUT WILL CHANGE THE HDRS AND OTHER PERCENTILES. A
LARGER M WILL SHORTEN THE HDR INTERVALS, AND A SMALLER M WILL
LENGTHEN THEM.

IF YOU WANT TO CHANGE M TYPE THE NEW VALUE (GREATER THAN 3).
IF YOU DO NOT WANT TO CHANGE M TYPE '0'.
? 20

HERE ARE SOME OF THE CHARACTERISTICS OF THE INVERSE CHI
DISTRIBUTION YOU ARE NOW CONSIDERING.

HYPOTHETICAL SAMPLE SIZE(M)	20.00	
10TH PERCENTILE	14.81	
25TH PERCENTILE	16.21	
50TH PERCENTILE	18.04	
75TH PERCENTILE	20.25	
90TH PERCENTILE	22.64	
50% HDR	15.51 TO	19.40
75% HDR	14.44 TO	21.17
95% HDR	12.89 TO	24.85

IF YOU WANT TO CHANGE M TYPE NEW VALUE ELSE '0'.? 22

HERE ARE SOME OF THE CHARACTERISTICS OF THE INVERSE CHI
DISTRIBUTION YOU ARE NOW CONSIDERING.

HYPOTHETICAL SAMPLE SIZE(M)	22.00	
10TH PERCENTILE	14.95	
25TH PERCENTILE	16.29	
50TH PERCENTILE	18.04	
75TH PERCENTILE	20.12	
90TH PERCENTILE	22.36	
50% HDR	15.65 TO	19.36
75% HDR	14.62 TO	21.03
95% HDR	13.11 TO	24.45

IF YOU WANT TO CHANGE M TYPE NEW VALUE ELSE '0'.? 24

(continued)

EXHIBIT 6.1 Continued

HERE ARE SOME OF THE CHARACTERISTICS OF THE INVERSE CHI
DISTRIBUTION YOU ARE NOW CONSIDERING.

[5]

HYPOTHETICAL SAMPLE SIZE(M)	24.00	
10TH PERCENTILE	15.08	
25TH PERCENTILE	16.36	
50TH PERCENTILE	18.04	
75TH PERCENTILE	20.02	
90TH PERCENTILE	22.13	
50% HDR	15.78 TO	19.32
75% HDR	14.78 TO	20.90
95% HDR	13.31 TO	24.11

IF YOU WANT TO CHANGE M TYPE NEW VALUE ELSE '0'.? 0

YOU CAN CHANGE THE CENTERING OF THE DISTRIBUTION BY
SPECIFYING A DIFFERENT MEDIAN. THIS WILL NOT AFFECT
THE HYPOTHETICAL SAMPLE SIZE.

[6]

IF YOU WANT TO CHANGE THE MEDIAN TYPE THE NEW VALUE.
IF YOU DO NOT TYPE '0'.
? 0

HERE ARE SOME OF THE CHARACTERISTICS OF THE INVERSE CHI
DISTRIBUTION FITTED TO YOUR PRIOR BELIEFS ABOUT SIGMA.

[7]

HYPOTHETICAL SAMPLE SIZE(M)	24.00	
DEGREES OF FREEDOM	23.00	
SCALE PARAMETER	85.26	
MODE	17.40	
10TH PERCENTILE	15.08	
25TH PERCENTILE	16.36	
50TH PERCENTILE	18.04	
75TH PERCENTILE	20.02	
90TH PERCENTILE	22.13	
50% HDR	15.78 TO	19.32
75% HDR	14.78 TO	20.90
95% HDR	13.31 TO	24.11

THIS COMPLETES THE SPECIFICATION OF A PRIOR DISTRIBUTION
ON SIGMA. IF YOU DO NOT WANT TO FIT A PRIOR DISTRIBUTION
ON THE MEAN YOU SHOULD RECORD THE PARAMETERS OF YOUR
PRIOR DISTRIBUTION ON SIGMA (DEGREES AND SCALE).

SOURCE: Output generated from CADA (Novick et al., 1983).

proceeds. The numbers in brackets on the exhibit identify points at
which some comment is made.

The program begins by asking the user to specify the 50th percentile
[1]. This is the value for which the user believes the standard deviation is
as likely to be above as below. The user should therefore accept even
odds on a bet that $\phi^{1/2}$ is below this value. Referring back to Table 6.4, we

see that the prior median value of $\phi^{1/2}$ was 17.62, when the posterior from the analysis of the first sample was used as a data-based prior. The investigator feels that the standard deviation is likely to be somewhat higher than this due to increased heterogeneity in the second sample, and therefore chooses 18.0 as the median value of his prior distribution for $\phi^{1/2}$.

The user is then asked to specify a 25th percentile and a 75th percentile [2]. CADA fits four possible distributions to these specifications. The investigator examines the percentiles of these distributions and selects one that best corresponds to his beliefs [3]. Some features of that distribution are then displayed [4]. The hypothetical sample size indicates that this prior distribution is the same as one that would have been obtained from an analysis of a previous sample of 29 observations. This seems a little high to the investigator because it means that he has almost as much information as the previous sample provided. He tries some other values and, after considering the percentiles, settles on the distribution for which the hypothetical sample size is 24 [5]. The user has the option of shifting the whole distribution up or down by changing the median [6], but, in this case, is satisfied with the fit and does not change the median. The characteristics of the final marginal distribution on $\phi^{1/2}$ are then displayed [7]. This distribution is $\phi^{1/2} \sim \chi^{-1}(23, 85.26)$. The investigator's prior beliefs about $\phi^{1/2}$ have been equated to a hypothetical prior sample of 24 observations for which R = 85.26.

At this point CADA assists the user in fitting a normal distribution to his prior beliefs about θ, conditional upon the median value of $\phi^{1/2}$, which is 18.04 in this problem. This fitting process is illustrated in Exhibit 6.2.

Again, the user is asked to specify a 50th percentile for his prior on [1]. The investigator does not have any reason for expecting it to be higher or lower than the mean in the previous sample, and chooses 255.81. Following this, he specifies a 25th and a 90th percentile. He chooses among four distributions fit to these percentiles [2], and examines the characteristics of the selected distribution [3]. After considering this, he decides to modify it slightly by decreasing the hypothetical sample size from 13.95 to 12 [4]. The investigator's beliefs about θ have been equated to a hypothetical prior sample of 12 observations with a mean of 255.81. The marginal t distribution on θ is then determined from the joint distribution [5]. It is $\theta \sim t(23, 255.81, 605.74)$. Note that the hypothetical sample size for the conditional distribution of the mean is less than that for the standard deviation. There is no reason that they have to be the same in non-data-based priors. If the investigator felt that

EXHIBIT 6.2

Assessment of a Prior Conditional and Marginal Distribution for θ

PRIOR DISTRIBUTION ON THE MEAN

THIS MODULE WILL ASSIST YOU IN SPECIFYING A PRIOR DISTRIBUTION
ON THE MEAN OF A NORMAL DISTRIBUTION.

SUPPOSE THE POPULATION STANDARD DEVIATION IS   18.04.
SPECIFY THE 25TH, 50TH, AND 90TH PERCENTILES OF YOUR PRIOR
DISTRIBUTION ON THE POPULATION MEAN.

SPECIFY 50TH. YOUR ODDS ARE EVEN THAT THE MEAN IS LESS THAN THIS
VALUE.? 255.81   **[1]**

SPECIFY 25TH. YOUR BETTING ODDS ARE 3 TO 1 THAT THE MEAN IS MORE
THAN THIS VALUE.? 253

SPECIFY 90TH. YOUR ODDS ARE 9 TO 1 THAT THE MEAN IS LESS THAN THIS
VALUE.? 262

HERE ARE THE PERCENTILES OF FOUR NORMAL DISTRIBUTIONS FITTED
TO YOUR PERCENTILE SPECIFICATIONS.

	10TH	25TH	50TH	75TH	90TH
1	250.47	253.00	255.91	258.62	261.15
2	249.62	252.55	255.81	259.07	262.00
3	250.21	253.00	256.10	259.21	262.00
4	250.06	252.86	255.96	259.06	261.85

COMPARE THE PERCENTILES OF THESE DISTRIBUTIONS AND DECIDE
WHICH DISTRIBUTION MOST CLOSELY CORRESPONDS TO YOUR PRIOR
BELIEFS. YOU CAN EITHER TENTATIVELY ACCEPT ONE OF THESE
DISTRIBUTIONS OR RESPECIFY THE PERCENTILES.

IF YOU WANT ONE OF THE DISTRIBUTIONS TYPE ITS NUMBER.   **[2]**
IF YOU WANT TO RESPECIFY THE PERCENTILES TYPE '0'.
? 2

HERE ARE SOME OF THE CHARACTERISTICS OF THE NORMAL   **[3]**
DISTRIBUTION YOU ARE NOW CONSIDERING.   THIS IS A
CONDITIONAL DISTRIBUTION SINCE IT IS ASSUMED THAT
THE POPULATION STANDARD DEVIATION IS   18.04.

HYPOTHETICAL SAMPLE SIZE (M)	13.96
MEAN=MODE=MEDIAN	255.81
STANDARD DEVIATION	4.83
10TH PERCENTILE	249.62
25TH PERCENTILE	252.55
75TH PERCENTILE	259.07
90TH PERCENTILE	262.00

IF YOU DO NOT FEEL THAT THIS VALUE OF M REFLECTS YOUR
PRIOR INFORMATION ABOUT THE MEAN YOU CAN SPECIFY A
DIFFERENT M. A SMALLER M WILL GIVE LONGER INTERVALS
AND A LARGER M SHORTER INTERVALS.

EXHIBIT 6.2 Continued

IF YOU WANT TO CHANGE M TYPE THE NEW VALUE ELSE '0'.? 12   **[4]**

HERE ARE SOME OF THE CHARACTERISTICS OF THE NORMAL
DISTRIBUTION YOU ARE NOW CONSIDERING.

```
HYPOTHETICAL SAMPLE SIZE (M)     12.00
MEAN=MODE=MEDIAN                 255.81
STANDARD DEVIATION                5.21
10TH PERCENTILE                 249.14
25TH PERCENTILE                 252.30
75TH PERCENTILE                 259.32
90TH PERCENTILE                 262.48
```

IF YOU WANT TO CHANGE M TYPE THE NEW VALUE ELSE '0'.? 0

YOU CAN CHANGE THE CENTERING OF THE DISTRIBUTION BY
SPECIFYING A DIFFERENT MEDIAN. THIS WILL NOT AFFECT
THE HYPOTHETICAL SAMPLE SIZE.

IF YOU WANT TO SPECIFY A DIFFERENT MEDIAN TYPE '1'.
IF YOU DO NOT TYPE '0'.
? 0

HERE ARE SOME OF THE CHARACTERISTICS OF THE PRIOR MARGINAL   **[5]**
DISTRIBUTION ON THE MEAN.

               STUDENT'S T DISTRIBUTION

```
DEGREES OF FREEDOM              23.00
SCALE PARAMETER                605.74
MEAN=MODE=MEDIAN               255.81
STANDARD DEVIATION               5.37
50% HDR               252.29 TO     259.33
75% HDR               249.75 TO     261.87
95% HDR               245.19 TO     266.43
```

WHEN YOU ARE READY TO CONTINUE   TYPE '1'? 1

THIS COMPLETES THE SPECIFICATION OF PRIOR DISTRIBUTION, YOU
MAY WISH TO RECORD THE FOLLOWING NUMBERS FOR LATER ANALYSIS.

THE PRIOR DISTRIBUTION ON THE STANDARD DEVIATION HAS AN
INVERSE CHI DISTRIBUTION WITH 23 DEGREES OF FREEDOM
AND THE SCALE PARAMETER 85.26.

THE PRIOR DISTRIBUTION ON THE MEAN HAS STUDENT'S T DISTRIBUTION
WITH MEAN 255.81 AND SCALE PRAAMETER 605.735

SOURCE: Output generated from CADA (Novick et al., 1983).

they should be the same, he could, of course, equate them. Changing these sample sizes would affect other characteristics of these distributions and the investigator would need to consider these changes carefully.

Although the marginal distribution of θ has the same mean as the data-based prior, it is more dispersed. The 95% HDR for the non-data-based prior is (245.19, 266.43) as compared to (249.53, 262.09) in the case of the data-based prior. This reflects the investigator's greater uncertainty about θ.

Given the joint prior just assessed, and the sample sufficient statistics $n = 27$, $\bar{x} = 259.78$ and $S^2 = 6937.94$ as described in Example 6-2, the formulas in Table 6.3 (with the substitution of $m_S = 23$ and $m_M = 12$ for m, where appropriate) yield the marginal distributions: $\phi^{1/2} \sim \chi^{-1}(50, 119.74)$ and $\theta \sim t(50, 258.56, 367.64)$. The 95% HDR for $\phi^{1/2}$ runs from 13.93 to 20.72, and the 95% HDR for θ runs from 253.11 to 264.01. It is interesting to compare these results with those obtained using the data-based prior, shown in Table 6.4. It can be seen that the posterior HDRs obtained here are longer than those in Table 6.4, and that the posterior mean (median) of θ is closer to the sample mean than it is in Table 6.4. This all reflects the investigator's greater prior uncertainty in the case of the non-data-based prior.

COMPARISON OF NORMAL MEANS AND VARIANCES

The methods that have been discussed extend naturally to the comparison of normal means and variances. Comparative methods are important in evaluation—the prototype of many evaluation studies involves the comparison of treatment and control groups. We will consider methods for comparing means in the case where variances are considered to be equal, and in the case where they are unequal. Following this, we will consider comparison of variances.

COMPARISON OF NORMAL MEANS WITH VARIANCES EQUAL

Analysis with a Noninformative Prior

We are interested here in the difference between the means θ_1 and θ_2 of two normal processes or populations with common unknown variance ϕ. As before, we let $\delta = \theta_2 - \theta_1$. Suppose that prior information about θ_1, θ_2, and ϕ is vague in relation to that contained in the sample. We will assume that our uncertainty about any of these parameters does not affect our uncertainty about any of the others. The joint prior will then be the product of the individual noninformative priors on these parameters. We will take $p(\theta_1)$ and $p(\theta_2)$ to be uniform and $p(\phi)$ to be log-uniform.

The steps in combining the prior and the likelihood to obtain the posterior are similar to those discussed earlier in this chapter, although the problem is somewhat more complex because we must deal with joint functions of three variables rather than two. Here we will simply present the results. Suppose that we obtain two independent random samples with sufficient statistics n_1, $\bar{x}_1$, S_1^2, and n_2, $\bar{x}_2$, S_2^2, for samples 1 and 2, respectively. Then the joint posterior distribution of these three parameters is such that the marginal distribution of $\phi^{1/2}$ is

$$\phi^{1/2} \sim \chi^{-1}[n_1 + n_2 - 2, (S_1^2 + S_2^2)^{1/2}]$$

Conditional upon ϕ, the parameters θ_1, θ_2, and δ have normal distributions as discussed in Chapter 5. The marginal unconditional distribution of θ_1, which we obtain by integrating out ϕ and θ_2 is

$$\theta_1 \sim t[n_1 + n_2 - 2, \bar{x}_1, n_1^{-1}(S_1^2 + S_2^2)]$$

Likewise, the marginal distribution of θ_2, obtained by integrating out $\phi^{1/2}$ and θ_1 is

$$\theta_2 \sim t[n_1 + n_2 - 2, \bar{x}_2, n_2^{-1}(S_1^2 + S_2^2)]$$

The marginal unconditional distribution of δ is

$$\delta \sim t[n_1 + n_2^{-2}, \bar{x}_2 - \bar{x}_1, (n_1^{-1} + n_2^{-1})(S_1^2 + S_2^2)]$$

This means that

$$t = \frac{\delta - (\bar{x}_2 - \bar{x}_1)}{[(n_1^{-1} + n_2^{-1})(S_1^2 + S_2^2)/(n_1 + n_2 - 2)]^{1/2}}$$

has a standard t distribution on $n_1 + n_2 - 2$ degrees of freedom. This is similar to the sampling distribution of $(\bar{x}_2 - \bar{x}_1)$ used in the classical test for the difference between two means. In the Bayesian approach, how-

TABLE 6.5
Inference for Two Normal Means with a
Noninformative Prior Distribution

Prior Marginal Distributions:
$\phi^{1/2} \sim$ log-uniform
$\theta_1 \sim$ uniform
$\theta_2 \sim$ uniform
$(\theta_1, \theta_2, \phi^{1/2}$ independent$)$

Sample Sufficient Statistics:
$n_1, \bar{x}_1, S_1^2$
$n_2, \bar{x}_2, S_2^2$
(samples independent)

Posterior Marginal Distributions:
$\phi^{1/2} \sim \chi^{-1} [n_1 + n_2 - 2, (S_1^2 + S_2^2)^{1/2}]$
$\theta_1 \sim t [n_1 + n_2 - 2, \bar{x}_1, n_1^{-1} (S_1^2 + S_2^2)]$
$\theta_2 \sim t [n_1 + n_2 - 2, \bar{x}_2, n_2^{-1} (S_1^2 + S_2^2)]$
$\delta \sim t [n_1 + n_2 - 2, \bar{x}_2 - \bar{x}_1, (n_1^{-1} + n_2^{-1}) (S_1^2 + S_2^2)]$

ever, $(\bar{x}_2 - \bar{x}_1)$ is fixed and it is δ that is random. These results are summarized in Table 6.5.

We point out again that the separate marginal distributions are presented to aid in interpreting the posterior distribution, and to illustrate the changes from prior to posterior. They are not independent, however, and one cannot make joint inferences by simply multiplying probabilities obtained from marginal distributions. We will focus on inferences from the single marginal distribution of δ, and the issue of joint inference does not arise. CADA contains procedures for making joint inferences from joint posterior distributions when this is desired.

* * * * * *

EXAMPLE 6-4. AN ANALYSIS USING A NONINFORMATIVE PRIOR: EVALUATING INSTRUCTIONAL METHODS

An experiment is conducted to evaluate a new method of teaching mathematics to high school sophomores. Twenty-five students are randomly assigned to each of two groups. Group 1 is a control group receiving the standard method of instruction; Group 2 is an experimental group receiving the new method of instruction. At the completion of a unit of the material, a 100-point test is administered to assess the students' mastery of the subject. The investigator decides to analyze the data using noninformative priors. For reasons unrelated to the methods of instruction, the control group had attrition of 3 students and the experimental group had attrition of 1 student.

The following test results are obtained:

$$n_1 = 22 \qquad \bar{x}_1 = 62.27 \qquad s_{n1} = 11.30 \qquad (S_1^2 = 2809.18)$$

$$n_2 = 24 \qquad \bar{x}_2 = 76.22 \qquad s_{n2} = 10.97 \qquad (S_2^2 = 2888.18)$$

Substituting these values in the formulas in Table 6.1, the investigator obtains the following marginal distributions:

$$\phi^{\frac{1}{2}} \sim \chi^{-1}(44, 75.48)$$

$$\theta_1 \sim t(44, 62.27, 258.97)$$

$$\theta_2 \sim t(44, 76.22, 237.39)$$

$$\delta \sim t(44, 13.95, 496.36)$$

The posterior distribution of δ is shown in Figure 6.3. We can see that the distribution is centered well above zero.

Suppose that the investigator feels that the difference between the two groups must be at least 10 points to be of any practical significance. The probability that δ exceeds 10 is

$$P(\delta > 10.00) = P\left\{t > \left[\frac{10.00 - 13.95}{(496.36/44)^{\frac{1}{2}}}\right]\right\} = P(t > -1.17)$$

where t is a standardized t variate. For a standardized t distribution with 44 degrees of freedom, this probability is .88. The odds are over seven to one in favor of the difference being larger than 10 points.

The probability that θ_2 exceeds θ_1 by any amount is over .999, but this is not the relevant probability here. Unfortunately, this is often the only question considered in analyzing data such as these with hypothesis tests. The null hypothesis of no difference would be rejected with

```
OPTION 6:  GRAPH OF THE DENSITY FUNCTION   OVER 99% HDR
               STUDENT'S T DISTRIBUTION
DEGREES OF FREEDOM =     44.00                    MEAN =      13.95
SCALE PARAMETER    =    496.36   STANDARD DEVIATION =       3.44
-------------------------------------------------------------------
THESE ARE THE PARAMETERS OF THE DISTRIBUTION TO BE GRAPHED.
WHEN YOU ARE READY FOR THE GRAPH TO BE DISPLAYED TYPE '1'? 1

  DF=    44.00   MEAN=      13.95   ST.DEV=      3.44
    4.91 I\
    5.81 I\\
    6.72 I\\\\\
    7.62 I\\\\\\\\
    8.52 I\\\\\\\\\\I\\\
    9.43 I\\\\\\\\\\I\\\\\\\\\I
   10.33 I\\\\\\\\\\I\\\\\\\\\\I\\\\\\\
   11.24 I\\\\\\\\\\I\\\\\\\\\\I\\\\\\\\\\I\\\\\
   12.14 I\\\\\\\\\\I\\\\\\\\\\I\\\\\\\\\\I\\\\\\\\\\I\\\
   13.05 I\\\\\\\\\\I\\\\\\\\\\I\\\\\\\\\\I\\\\\\\\\\I\\\\\\\\\
   13.95 I>>>>>>>>>1>>>>>>>>>2>>>>>>>>>3>>>>>>>>>4>>>>>>>>>5
   14.85 I//////////I//////////I//////////I//////////I//////////
   15.76 I//////////I//////////I//////////I//////////I////
   16.66 I//////////I//////////I//////////I//////
   17.57 I//////////I//////////I///////
   18.47 I//////////I//////////I
   19.38 I//////////I///
   20.28 I/////////
   21.18 I//////
   22.09 I//
   22.99 I/
```

SOURCE: Output generated from CADA (Novick et al., 1980b).

Figure 6.3 Posterior Distribution of the Difference Between Group Means

$p < .001$; however, this is not the answer to the question that the investigator is asking.

* * * * * *

Analysis with an Informative Prior

Prior information can be incorporated in a manner similar to that in the case of the single mean. Suppose that we have prior samples with sufficient statistics m_1, $\overline{w}_1$, R_1^2 and m_2, $\overline{w}_2$, and R_2^2. Analyzing these data with a noninformative prior, the joint posterior will be such that the parameters have the marginal distributions shown at the top of Table 6.6. Using this joint distribution as a prior in the analyses of subsequent samples yields a joint posterior with the marginal distributions shown at the bottom of Table 6.6.

Although the notation is more complex here, the posterior parameters contain many of the same features that we have already seen in

TABLE 6.6
Inference for Two Normal Means with an
Informative Prior Distribution

Prior Marginal Distributions:

$\phi^{1/2} \sim \chi^{-1}[m_1 + m_2 - 2, (R_1^2 + R_2^2)^{1/2}]$

$\theta_1 \sim t[m_1 + m_2 - 2, \bar{w}_1, m_1^{-1}(R_1^2 + R_2^2)]$

$\theta_2 \sim t[m_1 + m_2 - 2, \bar{w}_2, m_2^{-1}(R_1^2 + R_2^2)]$

Sample Sufficient Statistics:

$n_1, \bar{x}_1, S_1^2$

$n_2, \bar{x}_2, S_2^2$

(samples independent)

Posterior Marginal Distributions:

$\phi^{1/2} \sim \chi^{-1}(\nu, \lambda^{1/2})$

$\theta_1 \sim t[\nu, \dfrac{m_1 \bar{w}_1 + n_1 \bar{x}_1}{m_1 + n_1}, (m_1 + n_1)^{-1} \lambda]$

$\theta_2 \sim t[\nu, \dfrac{m_2 \bar{w}_2 + n_2 \bar{x}_2}{m_2 + n_2}, (m_2 + n_2)^{-1} \lambda]$

$\delta \sim t\left[\nu, \dfrac{m_2 \bar{w}_2 + n_2 \bar{x}_2}{m_2 + n_2} - \dfrac{m_1 \bar{w}_1 + n_1 \bar{x}_1}{m_1 + n_1}, [(m_1 + n_1)^{-1} + (m_2 + n_2)^{-1}] \lambda\right]$

where

$\nu = m_1 + m_2 + n_1 + n_2 - 2$

$\lambda = R_1^2 + S_1^2 + \dfrac{m_1 n_1 (\bar{x}_1 - \bar{w}_1)^2}{m_1 + n_1} + R_2^2 + S_2^2 + \dfrac{m_2 n_2 (\bar{x}_2 - \bar{w}_2)^2}{m_2 + n_2}$

Table 6.3. Looking at the posterior distribution of δ, we see that the degrees of freedom parameter is based on the number of prior and sample observations in both groups. The mean is the difference between the weighted averages of the prior and sample means within each group. The scale factor is based on the prior and sample sums of squares for each group, plus a between prior and sample sum of squares for each group.

* * * * * *

EXAMPLE 6-5. AN ANALYSIS USING AN INFORMATIVE PRIOR: EVALUATING INSTRUCTIONAL METHODS

The evaluation of methods of instruction is repeated the following year. The methods of instruction and the characteristics of the students are considered to be the same as the previous year, and the posterior from the previous year is used as the prior for the subsequent analysis. This posterior is in the form required for the prior in Table 8.2. The relevant prior quantities are

$$m_1 = 22 \qquad \overline{w}_1 = 62.27 \qquad R_1^2 = 2809.18$$

$$m_2 = 24 \qquad \overline{w}_2 = 76.22 \qquad R_2^2 = 2888.18$$

Students are again randomly assigned to the two conditions. The test is administered at the completion of the unit, and the sample sufficient statistics are:

$$n_1 = 23 \qquad \overline{x}_1 = 59.93 \qquad s_{n1} = 11.73 \qquad (S_1^2 = 3164.64)$$

$$n_2 = 22 \qquad \overline{x}_2 = 75.03 \qquad s_{n2} = 11.21 \qquad (S_2^2 = 2764.61)$$

Substituting these values in the formulas for the posterior parameters we obtain the following marginal distributions:

$$\phi^{\frac{1}{2}} \sim \chi^{-1}(89, 108.19)$$

$$\theta_1 \sim t(89, 61.07, 260.10)$$

$$\theta_2 \sim t(89, 75.65, 254.44)$$

$$\delta \sim t(89, 14.58, 514.54).$$

All of these distributions are more concentrated than those obtained in Example 6-4. In comparing this to the previous results, we see that this second set of data has shifted the posterior distribution up, from a mean of 13.95 to 14.58. In addition, consideration of both sets of data has resulted in a more precise picture of the size of δ—the 99% HDR is now (8.25, 20.91) rather than (4.91, 22.99) as it was when only the first set of data was considered. From the posterior distribution of δ it can be determined that the probability that δ exceeds 10 points is now. 97. The odds are over 32 to 1 that $\delta > 10.00$.

It is interesting to note that had the investigator conducted classical tests of the hypotheses H_0: ≤ 10.00 and H_1: > 10.00 on the first and second sets of data separately, the investigator would have obtained a p-value of .12 in the first study and a p-value of .08 in the second study. The first would be judged nonsignificant at conventional levels, and the second might be judged significant if the investigator were willing to accept a significance level of .10. At best, the investigator would have one nonsignificant and one marginally significant result, and would probably be somewhat reluctant to conclude that the difference is greater than ten points. Yet the Bayesian analysis shows that with the combined samples $P(\delta > 10) = .97$, and the probability that $\delta > 10$ is therefore over thirty times greater than the probability that $\delta \leq 10$. One can see the problem in conducting hypothesis tests with arbitrary significance levels and then using the number of significant results as a measure of evidence against the null hypothesis.

* * * * * *

COMPARISON OF NORMAL MEANS
WITH VARIANCES UNEQUAL

The analyses outlined in Tables 6.5 and 6.6 involve the assumption that the variances of the data-generating processes for the two groups are equal, $\phi_1 = \phi_2 = \phi$. This assumption will not be appropriate for some problems. We now consider how this is handled in the Bayesian approach.

If the prior distributions for each group are independent and the samples are independent, the posterior distributions for each group will also be independent. The posterior marginal distribution of δ obtained by integrating out ϕ_1 and ϕ_2 is now a Behrens distribution rather than a t distribution. Like the t distribution, and the normal distribution, the Behrens distribution is bell shaped and symmetrical. This distribution is more complicated—it depends upon five parameters.

Suppose that with uniform priors, we obtain the following posterior marginal distributions on θ_1 and θ_2: $\theta_1 \sim t(n_1, \bar{x}_1, n_1^{-1}S_1^2)$ and $\theta_2 \sim t(n_2, \bar{x}_2, n_2^{-1}S_2^2)$. Then the posterior Behrens distribution on δ is

$$
\delta \sim B \left\{ n_1 - 1, n_2 - 1, \tan^{-1} \left[\frac{(n_1 - 1)^{-1}(n_1^{-1}S_1^2)}{(n_2 - 1)^{-1}(n_2^{-1}S_2^2)} \right]^{\frac{1}{2}}, \bar{x}_2 - \bar{x}_1, \right.
$$
$$
\left. [(n_1 - 1)^{-1}(n_1^{-1}S_1^2) + (n_2 - 1)^{-1}(n_2^{-1}S_2^2)]^{\frac{1}{2}} \right\}
$$

The first two parameters are degrees of freedom from groups 1 and 2. The third parameter relates to the ratio of the two variances. This is actually an angle: $\tan^{-1}$ means the angle that has as its tangent the quantity in brackets. The fourth parameter is the mean of the distribution, and the fifth is a scale factor. The nature of this distribution can be more easily seen in its standardized form. The quantity

$$
B = \frac{\delta - (\bar{x}_2 - \bar{x}_1)}{[(n_1^{-1}S_1^2)/(n_1 - 1) + (n_2^{-1}S_2^2)/(n_2 - 1)]^{\frac{1}{2}}}
$$

is said to have a standard Behrens distribution on $n_1 - 1$ and $n_2 - 1$ degrees of freedom with angle ψ, as described above.

* * * * * *

EXAMPLE 6-6. AN ANALYSIS OF CLIENT SATISFACTION DATA

A client satisfaction questionnaire is administered to a sample of persons receiving outpatient services at a community mental health center. The responses of males and females receiving group therapy are examined for differences on a particular item. This item is an 11-point Likert-type scale, running from 1-"very dissatisfied" to 11-"very satisfied." The investigator does not assume that the variances will be equal. The investigator's prior information is vague, and he or she decides to analyze the data with independent uniform priors on θ_1 and θ_2, and log-uniform priors on ϕ_1 and ϕ_2. Data were obtained independently from 13 males with $\bar{x}_1 = 7.43$ and $s_{n1} = 2.30$, and from 18 females with $\bar{x}_2 = 7.81$ and $s_{n2} = 1.24$. The sample sufficient statistics are therefore

$$n_1 = 13 \qquad \bar{x}_1 = 7.43 \qquad S_1^2 = 68.77$$

$$n_2 = 18 \qquad \bar{x}_2 = 7.81 \qquad S_2^2 = 27.68$$

The independent posterior distributions of θ_1 and θ_2 are $\theta_1 \sim t(12, 7.43, 5.29)$ and $\theta_2 \sim t(17, 7.81, 1.54)$. Substituting these values in the formulas for the parameters shown above, we obtain

$$\delta \sim B(12, 17, 65.6°, .38, .73)$$

where the angle 65.6° was found using an arctangent function on a pocket calculator.

Tables for evaluating the Behrens distribution are available (Novick and Jackson, 1974; Fisher and Yates, 1964). However, it is much easier to evaluate this distribution using CADA. In fact, the user can obtain the parameters of the posterior Behrens distribution simply by entering the parameters of the separate posterior distributions for θ_1 and θ_2. Various features of this distribution could be reported. The 95% HDR, for example, is (−1.20, 1.96). The probability that females are more satisfied than males on this item is $P(\delta > 0) = .69$.

* * * * * *

The analysis can be just as easily carried out with informative priors. As long as prior information about the two groups is independent, and the samples are independent, then the posteriors are independent. The parameters of these posterior distributions could be entered into the formulas for the parameters of the posterior Behrens distribution.

The classical analysis of this problem has been a source of controversy. Analysis within the classical framework is made difficult by the fact that the sampling distribution of $\bar{x}_2 - \bar{x}_1$ depends upon the ratio σ_1^2 / σ_2^2, which is unknown. Consequently, it is not possible to give exact probabilities of Type I error, although approximations are available. This problem does not arise in the Bayesian analysis because σ_1 and σ_2

are random and are eliminated from the posterior marginal distribution of δ by integration. The classical results will not agree numerically with the Bayesian results as they did for other problems that were discussed. Using the Satterthwaite approximation described in Winer (1971), the boundaries of the classical 95% confidence interval are determined to be -1.16 and 1.92. These differ from the boundaries of the 95% HDR described above.

Fisher provided an alternative approach to this problem based on his theory of *fiducial inference*. Although he worked within the framework of relative frequency notions of probability, he arrived at a distribution for the parameter δ that corresponds to the Bayesian posterior just discussed. In fact, this is known as the Behrens-Fisher problem in the statistical literature. There has been much discussion of the validity of Fisher's argument and it has not been widely accepted. Bayesians can reach the same result using standard Bayesian methods, and other classical theorists disagree with the argument that leads to a probability distribution for a parameter. Robinson (1982) summarizes aspects of this controversy.

COMPARISON OF VARIANCES

In comparing two groups, it may be useful to compare the variances as well as the means. For example, we may be interested in how the variation in outcome within a treatment group compares with that within the control group. This could be important in judging the merits of the treatment. In addition, we may wish to examine the homogeneity assumption that $\phi_1 = \phi_2 = \phi$, which underlies many of the procedures discussed above.

We will compare normal variances in the case where the means are unknown and unequal. Analyzing each group separately, the posterior marginal distributions of the variances are $\phi_1 \sim \chi^{-2}(\nu_1, \lambda_1)$ and $\phi_2 \sim \chi^{-2}(\nu_2, \lambda_2)$. The two variances are usually compared by examining their ratio. The quantity

$$F = \frac{\phi_2 (\lambda_2/\nu_2)^{-1}}{\phi_1 (\lambda_1/\nu_1)^{-1}}$$

has a standard F distribution with ν_1 and ν_2 degrees of freedom, and the similarity to the F statistic discussed in Chapter 2 can be seen.

We will be primarily interested in the question of whether ϕ_2/ϕ_1 exceeds same value k. For example, if we want to know whether $\phi_2 > \phi_1$, then we are interested in whether $\phi_2/\phi_1 > 1$. If we wanted to know whether ϕ_2 was twice as large as ϕ_1, then the relevant value of k would be 2. The probability that $\phi_2/\phi_1 > k$ is

$$P\left[F > k \; \frac{(\lambda_2/\nu_2)^{-1}}{(\lambda_1/\nu_1)^{-1}}\right]$$

Given the values of the scale parameters and degrees of freedom of ϕ_1 and ϕ_2, we can calculate F and evaluate this using tables for the F distribution. This requires care in keeping the degrees of freedom in the right order, and often involves interpolation. We will instead use CADA in making this comparison.

Using CADA, one compares the standard deviations $\phi_1^{1/2}$ and $\phi_2^{1/2}$ rather than the variances; however, the principles remain the same. Returning to Example 6-5, the sample standard deviations are $s_{n1} = 2.30$ and $s_{n2} = 1.24$. The investigator wants to determine the probability that $\phi_1^{1/2} > \phi_2^{1/2}$. The posterior distributions of the standard deviations are $\phi_1^{1/2} \sim \chi^{-1}(12, 8.29)$ and $\phi_2^{1/2} \sim \chi^{-2}(17, 5.26)$. Entering the values of the parameters and the value of k = 1 into CADA, the investigator finds that the probability that $\phi_1^{1/2}$ exceeds $\phi_2^{1/2}$ is .98. Thus the use of the Behrens distribution in Example 6-5 is warranted.

FURTHER READING

Inference for the normal mean with unknown variance is discussed in Box and Tiao (1973), Hays (1973), Lindley (1965b), Phillips (1973), and Schmitt (1969). Novick and Jackson (1974) provide a thorough discussion, treating the assessment and incorporation of prior information in detail. Pitz (1982) describes another assessment procedure.

Comparisons of means and variances using noninformative prior distributions are discussed in Box and Tiao (1973), Lindley (1965b), Phillips (1973), and Schmitt (1969). Novick and Jackson (1974) discuss these topics using both noninformative and informative prior distributions. The analyses discussed in this chapter extend to comparisons of many means and Bayesian analysis of variance; an introduction to Bayesian analysis of variance is provided in Lewis (1982), and a full discussion is provided in Box and Tiao (1973).

NOTES

1. Press (1980, 1982) describes CADA along with other Bayesian computer programs. In contrast to CADA, most of the other programs are special-purpose programs. Another general purpose package not cited in these compilations is the Bayesian decision theory package described in Schlaifer (1971).

2. Tables of the inverse chi distribution can be found in Novick and Jackson. As these authors point out, percentage points of the inverse chi distribution can also be obtained from a table of the cumulative distribution of a chi-squared variate. The method is as follows:

$$P\% \text{ point of } \chi^{-1}(v, \lambda^{1/2}) = \left[\frac{\lambda}{(100 - P)\% \text{ point of } \chi^2(v)}\right]^{1/2}$$

To find the 95th percentile of $\chi^{-1}(20, 100)$, say, we begin by finding the $(100\text{th} - 95\text{th}) = 5\text{th}$ percentile of $\chi^2(20)$. Using a χ^2 table such as that provided in Fisher (1970: 112-113), we find this 5th percentile to be 10.85. The 95th percentile of $\chi^{-1}(20, 100)$ is therefore $(100/10.851)^{1/2} = 3.04$.

7

Inference in Linear Regression

Regression analysis is a flexible and widely used technique for analyzing data obtained in experimental and nonexperimental observational studies. The use of regression techniques contributes to understanding of the functional relationships among variables, and provides a basis for prediction and control. This latter aspect is important in evaluation and policy research, where the goal may be to identify variables that are amenable to manipulation in order to bring about some desired outcome.

In simple regression, we consider the relationship between an independent (or predictor) variable and a dependent (or criterion) variable; in multiple regression we look at the relationship between a set of independent variables and a dependent variable. We begin the chapter by considering simple regression. The related topic of inference for correlation coefficients is then examined. Finally, multiple regression is discussed.

SIMPLE LINEAR REGRESSION

THE REGRESSION MODEL

We are interested here in the relationship between a single independent variable X and a dependent variable Y. Suppose we have n paired observations (x, y) on X and Y. We assume the observations to be related linearly as follows:

$$y_i = \alpha + \beta x_i + e_i \qquad i = 1, 2, \ldots, n$$

where y_i is the ith observation on the dependent variable, x_i is the ith observation on the independent variable, and e_i is the ith unobserved value of the random error. The coefficients α and β are regression parameters—the intercept and slope of the regression line, respectively; α is the value of Y of the origin of X, and β represents the change in Y for a unit change in X. The y values are assumed to be randomly sampled; the x values may either be randomly sampled along with the y values or they may be preselected by the investigator. The principles of Bayesian regression are the same in both cases. In addition, we assume that the errors e are independent and that for fixed α, β, and x, e_i is normally distributed with mean 0 and common variance ϕ. It follows that the distribution of Y given x is normal with mean $\alpha + \beta x$ and residual variance ϕ, for all x. When X is random it is assumed that x and e are independent, and that the distribution p(x) does not involve the parameters α, β, or ϕ.

The density function for an observation on Y, conditional upon some value of x and the parameters α, β, and ϕ, is the normal function

$$p(y \mid x, \alpha, \beta, \phi) = (2\pi\phi)^{-\frac{1}{2}} \exp\left[\frac{[y - (\alpha + \beta x)]^2}{-2\phi} \right]$$

and the joint density of n independent observations on Y, given the x values (and α, β, and ϕ), is the product of the n individual densities.

$$p(y \mid x, \alpha, \beta, \phi) = (2\pi\phi)^{-n/2} \exp\left[\frac{\sum\limits_{i=1}^{n} (y_i - \alpha - \beta x_i)^2}{-2\phi} \right]$$

THE LIKELIHOOD FUNCTION

The likelihood function for the parameters given the observations is then

$$\ell(\alpha, \beta, \phi | y, x) \propto \phi^{-n/2} \exp\left[\frac{\sum\limits_{i=1}^{n} (y_i - \alpha - \beta x_i)^2}{-2\phi}\right]$$

Novick and Jackson (1974) and Lindley (1965a, 1965b) introduce a transformation here that simplifies the Bayesian analysis. Denote the ordinate at the mean of the x values as α^*. That is, $\alpha^* = \alpha + \beta\bar{x}$; we then have $\alpha = \alpha^* - \beta\bar{x}$. We are, in effect, treating $\bar{x}$ as the origin for the x values. Substituting for α in the summation term in the likelihood and rearranging, this term becomes

$$\sum_{i=1}^{n} (y_i - \alpha - \beta x_i)^2 = S^2 + n(\alpha^* - a^*)^2 + S_x^2(\beta - b)^2 \qquad [7.1]$$

where:

$$a^* = \bar{y} = n^{-1} \sum_{i=1}^{n} y_i \qquad\qquad b = \frac{S_{xy}}{S_x^2}$$

$$S_x^2 = \sum_{i=1}^{n} (x_i - \bar{x})^2 \qquad\qquad S_y^2 = \sum_{i=1}^{n} (y_i - \bar{y})^2$$

$$S_{xy} = \sum_{i=1}^{n} (x_i - \bar{x})(y_i - \bar{y}) \qquad\qquad S^2 = S_y^2 - \frac{(S_{xy})^2}{S_x^2}$$

The sample information enters the likelihood through the sufficient statistics n, $\bar{y}$, S_x^2, S_{xy}, and S^2. The left-hand side of equation 7.1 can be expressed using a^* and this becomes

$$\sum_{i=1}^{n} [y_i - \alpha^* - \beta(x_i - \bar{x})]^2$$

This transformation of the intercept in effect "centers" the independent variable at its mean. The slope parameter β remains the same in the centered and uncentered models; however, the intercept is different as discussed above.

Again, we have made no assumptions about the nature of the x's. It turns out that we arrive at the same likelihood function whether the values of x are selected by the investigator or randomly sampled jointly with the y values. Because the data enter Bayesian inference only through the likelihood, the form of the posterior distribution will be the same in either case.

ANALYSIS WITH A NONINFORMATIVE PRIOR

Our noninformative joint prior for $\alpha*$, β and ϕ will be the product of independent uniform priors on $\alpha*$, β, and $\ln \phi$. Thus we have the prior $P(\alpha*, \beta, \phi) \propto \phi^{-1}$. Multiplying the prior and the likelihood, we obtain the posterior distribution

$$p(\alpha*, \beta, \phi | y, x) \sim \phi^{-(n/2)-1} \exp \left[\frac{S^2 + n(\alpha* - a*)^2 + S_x^2(\beta - b)^2}{-2\phi} \right] \quad [7.2]$$

The marginal distribution of the variance ϕ, obtained by integrating out $\alpha*$ and β, is $\phi \sim \chi^{-2}(n - 2, S^2)$. When ϕ is known equation 7.2 reduces to

$$p(\alpha*, \beta | \phi, y, x) \sim \exp \left[\frac{n(\alpha* - a*)^2}{-2\phi} \right] \exp \left[\frac{S_x^2(\beta - b)^2}{-2\phi} \right]$$

We see that the posterior is the product of two terms, one containing $\alpha*$ and the other containing β. In addition, each term is a kernel of a normal density function. Therefore, given ϕ, the posterior distributions of $\alpha*$ and β are normal and they are independent. We have $\alpha* \sim N(\bar{y}, n^{-1}\phi)$ and $\beta \sim N(b, S_x^{-2}\phi)$, where $S_x^{-2} = (S_x^2)^{-1}$.

When ϕ is unknown, the marginal distributions of the intercept and slope are $\alpha* \sim t(n-2, \alpha*, n^{-1}S^2)$ and $\beta \sim t(n-2, b, S_x^{-2}S^2)$. These results are summarized in Table 7.1.

The advantage of transforming the intercept to $\alpha*$ is that $\alpha*$ and β are independent given ϕ, and are uncorrelated. This is not true of α and β and the likelihood and posterior involving α are somewhat more compli-

TABLE 7.1
**Inference for the Residual Variance, Intercept, and Slope
for the Normal Regression Model Using a
Noninformative Prior Distribution**

Prior Marginal Distributions:

$\phi \sim$ log-uniform
$\alpha^* \sim$ uniform
$\beta \sim$ uniform
(ϕ, α^*, β independent)

Sample Sufficient Statistics:

$n, \bar{y}, S_x^2, S_{xy}, S^2$

Posterior Marginal Distributions:

$\phi \sim \chi^{-2} (n - 2, S^2)$
$\alpha^* \sim t(n - 2, a^*, n^{-1} S^2)$
$\beta \sim t(n - 2, b, S_x^{-2} S^2)$

cated. A presentation based on the original intercept α can be found in Zellner (1971: 58-63).

Consider inference for the slope parameter β. In the Bayesian analysis

$$t = \frac{\beta - b}{[(S_x^{-2} S^2)/(n - 2)]^{1/2}} \qquad [7.3]$$

has a standard t distribution on $n - 2$ degrees of freedom, and this can be used in making various probability statements about the value of β.

In classical inference, ordinary least squares estimates of regression parameters are obtained by finding values that minimize the sum of squares of the residuals in the sample

$$\sum_{i-1}^{n} e_i^2 = \sum_{i-1}^{n} (y_i - \alpha - \beta x_i)^2 = \sum_{i=1}^{n} [y_i - \alpha^* - \beta(x_i - \bar{x})]^2$$

We can see from equation 7.1 that the residual sum of squares will be minimized when $\alpha^* = a^* = \bar{y}$ and $\beta = b = S_{xy}/S_x^2$. That is, the classical least squares estimators of α^* and β are a* and b, respectively. Thus with a noninformative prior, the means of the posterior distributions of α^* and β correspond with the classical estimates.

In the classical approach, the statistic

$$t = \frac{b - \beta}{[(S_x^{-2} S^2)/(n - 2)]^{\frac{1}{2}}}$$

has a standard t distribution on $n - 2$ degrees of freedom, and with a reversal in sign, this is numerically identical to equation 7.3. However, here the parameter β is fixed and the statistic b is random, whereas in the Bayesian analysis the statistic is fixed and the parameter is random.

The posterior distribution of α^* is the distribution of the intercept at $\bar{x}$, the mean of the x values. The intercept α_0 at some particular value x_0 is distributed

$$\alpha_0 \sim t \left\{ n - 2,\ a^* + b(x_0 - \bar{x}),\ \left[\frac{1}{n} + \frac{(x_0 - x)^2}{S_x^2}\right] S^2 \right\} \qquad [7.4]$$

This provides the basis for inference about the dependent variable of Y for any particular x-value of interest. Because the scale factor contains the quantity $(x_0 - \bar{x})^2$, the distribution of α_0 will be "tightest" when $x_0 = \bar{x}$. The further x_0 is from $\bar{x}$, the greater the uncertainty will be in the value of α_0. We can see that when $x_0 = \bar{x}$, this reduces to the posterior distribution for α^*.

When $x_0 = 0$, α_0 is the value of the intercept at the origin. Substituting 0 for x_0 in the formulas for the hyperparameters in the distribution 7.4, the distribution of the intercept at $x_0 = 0$ can be seen to be

$$\alpha \sim t \left[n - 2,\ a^* - b\,x,\ \left(\frac{1}{n} + \frac{\bar{x}^2}{S_x^2}\right) S^2 \right]$$

All of these posterior distributions coincide numerically with classical sampling distributions used in making inferences about α_0, but, again, the interpretations are very different.

In the classical framework, squared multiple correlation coefficients R^2 expressing percentage of variance explained receive considerable attention in the evaluation of the results. In the Bayesian framework, attention is directed toward posterior distributions of quantities of interest rather than R^2 statistics. The relationship between R^2 and posterior probabilities is considered in the final section of this chapter.

$$* \quad * \quad * \quad * \quad * \quad *$$

EXAMPLE 7-1. AN ANALYSIS USING A NONINFORMATIVE PRIOR: EVALUATING A HYPERTENSION REDUCTION PROGRAM

An investigator is examining the effectiveness of an exercise program in reducing blood pressure for persons with mild hypertension. Group exercise sessions are scheduled three times per week for ten weeks. A frequently occurring problem in evaluating a program is that persons within a treatment group may receive the treatment in varying degrees. If a measure of the extent of the treatment actually delivered to each individual is available, it can be useful to regress the outcome measure on the amount of treatment received to get some idea of the treatment-outcome relationship within the treatment group. In this study, there was substantial variation among individuals in attendance at the exercise sessions, and the investigator wishes to determine how this is related to the outcome.

There are 34 participants in the program. The average number of sessions attended is $\bar{x} = 25.36$ (out of a total of 30) with a standard deviation of $s_n = 2.30$. S_x^2 is 179.86. The average decrease in diastolic pressure is $\bar{y} = 6.38$ millimeters of mercury, with a standard deviation of $s_n = 3.12$. (Note that the decrease is discussed here as a positive quantity.) S_y^2 is 330.97, S_{xy} is 87.83, and $S^2 = S_y^2 - (S_{xy}^2/S_x^2)$ is determined to be 288.08.

Given the assumptions of the linear regression model, the investigator analyzes these data and obtains the following marginal posterior distributions:

$$\phi \sim \chi^{-2}(32, 288.08),$$

$$\alpha^* \sim t\left(32, 6.38, \frac{288.08}{34}\right)$$

```
OPTION 6:   GRAPH OF THE DENSITY FUNCTION   OVER 99% HDR
               STUDENT'S T DISTRIBUTION
DEGREES OF FREEDOM  =    32:00              MEAN =     0.49
SCALE PARAMETER     =    1.60   STANDARD DEVIATION =    0.23
---------------------------------------------------------------
THESE ARE THE PARAMETERS OF THE DISTRIBUTION TO BE GRAPHED.
WHEN YOU ARE READY FOR THE GRAPH TO BE DISPLAYED TYPE '1'? 1

   DF=    32.00   MEAN=       0.49   ST.DEV=      0.23
  -0.12 I\
  -0.06 I\\
   0.00 I\\\\
   0.06 I\\\\\\\\
   0.12 I\\\\\\\\\\\I\\\
   0.18 I\\\\\\\\\\I\\\\\\\\\
   0.25 I\\\\\\\\\\I\\\\\\\\\\\I\\\\\\\
   0.31 I\\\\\\\\\\I\\\\\\\\\\\I\\\\\\\\\\\I\\\\\
   0.37 I\\\\\\\\\\I\\\\\\\\\\\I\\\\\\\\\\\I\\\\\\\\\\\I\\
   0.43 I\\\\\\\\\\I\\\\\\\\\\\I\\\\\\\\\\\I\\\\\\\\\\\I\\\\\\\\\
   0.49 I>>>>>>>>>1>>>>>>>>>2>>>>>>>>>3>>>>>>>>>4>>>>>>>>>5
   0.55 I//////////I///////////I///////////I///////////I///////////
   0.61 I//////////I///////////I///////////I///////////I//
   0.67 I//////////I///////////I///////////I//////
   0.73 I//////////I///////////I///////
   0.80 I//////////I///////////
   0.86 I//////////I///
   0.92 I/////////
   0.98 I////
   1.04 I//
   1.10 I/
```

SOURCE: Output generated from CADA (Novick et al., 1980b).

Figure 7.1 Posterior Distribution of β

which reduces to $\alpha^* \sim t(32, 6.38, 8.47)$, and

$$\beta \sim t \left(32, \frac{87.83}{179.86}, \frac{288.08}{179.86}\right)$$

which reduces to $\beta \sim t(32, .49, 1.60)$.

We will first consider inference for the slope parameter β. The posterior distribution of β is shown in Figure 7.1. We can see that most of the distribution lies well above zero. The posterior probability that β exceeds zero is over .98. The 95% HDR is the interval (.03, .95).

What the investigator has at this point is a statistical relationship between amount of treatment received and decrease in diastolic pressure. This relationship is not necessarily causal. If these data are all that are obtained in the entire study, the investigator is in the position of drawing conclusions from a one-group pretest-postest design. There may be alternative plausible hypotheses that would account for the relationship. One obvious alternative hypothesis is that attendance may be associated with commitment to make changes in diet, smoking, and so on, and that these changes may account for some or all of the decrease in blood pressure. Additional data would be required to rule out this possible explanation. Careful attention to design is necessary in making causal inferences, whether one is using classical or Bayesian statistics. For purposes of this discussion, we will assume that the investigator is simply analyzing part of the data obtained in a more comprehensive study and that these rival hypotheses can be ruled out.

The distribution of α^* is not of particular interest here. Nor is the distribution of the intercept at x = 0 of interest. In this particular sample, no values of x below 17 were observed. Extrapolating beyond the range of observed x values can lead to serious errors. Although a relationship may be approximately linear within the range of values included in the study, it could very well be nonlinear outside of that range. The distribution of the intercept at x = 0 is $\alpha_0 \sim t(32, -6.00, 1038.55)$. This says that the average change in blood pressure for persons not taking part in the exercise program would be an increase of 6.00 mm. This is clearly a nonsensical result based on an unjustified extrapolation.

Assuming that the design of the full study is such that the relationship between decrease in blood pressure and exercise could be given a causal interpretation, the investigator might want to report the distribution of the intercept at x = 30. This would be the average decrease in blood pressure for those persons experiencing the full treatment of 30 exercise sessions. This is assuming, of course, that x = 30 is within, or very close to, the range of values included in the study. This distribution is

$$\alpha_0 \sim t\left(32, 6.38 + .49(30.00 - 25.36), \left[\left(\frac{1}{34} + \frac{(30.00 - 25.36)^2}{179.86}\right)288.08\right]\right)$$

which is $\alpha_0 \sim t(32, 8.64, 42.96)$. Note that the scale factor of 42.96 is larger than that of 8.47 for α^*; as we noted earlier, the posterior distribution of the intercept becomes more spread out as $x_0 - \bar{x}$ increases. This distribution is shown in Figure 7.2. The 95% HDR runs from 6.28 to 11.00.

*   *   *   *   *   *

```
OPTION 6:  GRAPH OF THE DENSITY FUNCTION   OVER 99% HDR
               STUDENT'S T DISTRIBUTION
DEGREES OF FREEDOM =    32.00                    MEAN =    8.64
SCALE PARAMETER    =    42.96  STANDARD DEVIATION =    1.20
----------------------------------------------------------------
THESE ARE THE PARAMETERS OF THE DISTRIBUTION TO BE GRAPHED.
WHEN YOU ARE READY FOR THE GRAPH TO BE DISPLAYED TYPE '1'? 1

  DF=    32.00   MEAN=       8.64   ST.DEV=       1.20
     5.47 I\
     5.78 I\\
     6.10 I\\\\
     6.42 I\\\\\\\\
     6.74 I\\\\\\\\\\I\\\
     7.05 I\\\\\\\\\\\I\\\\\\\\\
     7.37 I\\\\\\\\\\I\\\\\\\\\\\I\\\\\\\
     7.69 I\\\\\\\\\\I\\\\\\\\\\\\\I\\\\\\\\\\\\\\I\\\\\
     8.01 I\\\\\\\\\\\I\\\\\\\\\\\\I\\\\\\\\\\\\\\I\\\\\\\\\\\\\I\\
     8.32 I\\\\\\\\\\\I\\\\\\\\\\\\\I\\\\\\\\\\\\\\I\\\\\\\\\\\\\\I\\\\\\\\\
     8.64 I>>>>>>>>>1>>>>>>>>>>2>>>>>>>>>>3>>>>>>>>>>4>>>>>>>>>>5
     8.96 I//////////I//////////I//////////I//////////I/////////
     9.27 I//////////I//////////I//////////I//////////I//
     9.59 I//////////I//////////I//////////I/////
     9.91 I//////////I//////////I//////
    10.23 I//////////I/////////
    10.54 I//////////I///
    10.86 I////////
    11.18 I////
    11.50 I//
    11.81 I/
```

SOURCE: Output generated from CADA (Novick et al., 1970b).

Figure 7.2 Posterior Distribution of α_0 at $\alpha_0 = 30$

ANALYSIS WITH AN
INFORMATIVE PRIOR

The natural conjugate family of distributions for the simple linear regression problem is that for which ϕ has a marginal distribution that is χ^{-2}, and the regression parameters α^* and β have a bivariate normal distribution, given ϕ. The posterior distribution just discussed was of this form. When prior information is provided by an analysis of the previous sample, the posterior obtained from that previous sample will be a member of the natural conjugate family and in the appropriate form for use as an informative prior in a subsequent analysis.

Suppose that we obtained a previous sample with statistics n, $\bar{y}$, S_x^2, S_{xy}, S^2, and $\bar{x}$, relabeled as m, $\bar{v}$, R_w^2, R_{wv}, R^2, and $\bar{w}$ to distinguish them from the statistics to be obtained in a subsequent study.[1] Note that we are using R^2 here to denote the prior sum of squared deviations; it is not the classical R^2 statistic expressing percentage of variance explained.

Given the model and a noninformative prior distribution, the marginal posterior distributions are

$$\phi \sim \chi^{-2}(m-2, R^2)$$
$$\alpha^* \sim t(m-2, \bar{v}, m^{-1} R^2)$$
$$\beta \sim t(m-2, R_{xy}/R_x^2, R_x^{-2}R^2)$$

Now suppose that we obtain a second sample with statistics n, $\bar{y}$, S_x^2, S_{xy}, S^2 and $\bar{x}$. Using the posterior from the analysis of the first sample as a prior in the analysis of the second sample, we obtain the marginal posterior distributions shown in Table 7.2. The degrees of freedom parameters are based on the number of prior and sample observations, and the scale parameters reflect both sum of squared deviations within prior and sample and sum of squared deviations between prior and sample. Differences between the prior and sample add to the dispersion of these posterior distributions. Note that the posterior intercept is denoted α^{**}. This is the ordinate at the mean of the independent variable for the prior and sample observations combined; this mean is $(m\bar{w} + n\bar{x})/(m + n)$.

* * * * * *

EXAMPLE 7-2. AN ANALYSIS USING AN INFORMATIVE PRIOR: EVALUATING A HYPERTENSION REDUCTION PROGRAM

Following the study described in Example 7-1, the posterior distributions that were discussed represent the beliefs of someone with vague prior information. Suppose that additional data are obtained in a subsequent study. These new data can then be used to revise the posteriors from the previous example.

From the previous study the investigator has the following quantities:

m = 34	$\bar{w}$ = 25.36	$\bar{v}$ = 6.38

$R_w^2 = 179.86$ $\qquad$ $R_v^2 = 330.97$ $\qquad$ $R_{wv} = 87.83$ $\qquad$ $R^2 = 288.08$

TABLE 7.2
Inference for the Residual Variance, Intercept, and Slope
for the Normal Regression Model Using an
Informative Prior Distribution

Prior Marginal Distributions:

$\phi \sim \chi^{-2}(m-2, R^2)$

$a^* \sim t(m-2, \bar{v}, m^{-1} R^2)$

$\beta \sim t(m-2, R_{xy}/R_x^2, R_x^{-2} R^2)$

Sample Sufficient Statistics:

$n, \bar{y}, S_x^2, S_{xy}, S^2$

Posterior Marginal Distributions:

$\phi \sim \chi^{-2}(m+n-2, D)$

$a^{**} \sim t[m+n-2, a^{**}, (m+n)^{-1} D]$

$\beta \sim t(m+n-2, b, A^{-1} D)$

where

$a^{**} = \dfrac{m\bar{v} + n\bar{y}}{m+n}$

$b = C/A$

$A = R_w^2 + S_x^2 + \dfrac{mn}{m+n}(\bar{w} - \bar{x})^2$

$B = R_v^2 + S_y^2 + \dfrac{mn}{m+n}(\bar{v} - \bar{y})^2$

$C = R_{wv} + S_{xy} + \dfrac{mn}{m+n}(\bar{w} - \bar{x})(\bar{v} - \bar{y})$

$D = B - (C^2/A)$

Suppose that the following statistics are obtained in the subsequent study:

$n = 42$ $\bar{x} = 27.08$ $\bar{y} = 8.81$

$S_x^2 = 127.16$ $S_y^2 = 526.33$ $S_{xy} = 75.02$ $S^2 = 482.07$

Combining these quantities as indicated in Table 7.2, the investigator obtains these posterior marginal distributions:

$$\phi \sim \chi^{-2}(74, 807.56)$$

$$a^{**} \sim t(74, 7.72, 10.63)$$

$$\beta \sim t(74, .67, 2.23)$$

Obviously, doing this by hand can be tedious. These results can be obtained using CADA simply by entering the values of parameters of the prior distributions and sample statistics.

The distribution of belief about β is now more concentrated. The 95% HDR runs from .32 to 1.02; this is shorter than the 95% HDR of (.03, .95) obtained from the first study alone. The mean of this posterior distribution, .67, is greater than that of .49 obtained in the first study. In this case, the combined data indicate a stronger relationship between participation and decrease in blood pressure than the first set of data alone.

The posterior distribution of the intercept at x = 30 may be of interest here also. To simplify, we will let $\bar{x}_{mn} = (m\bar{w} + n\bar{x})/(m + n)$; This is the mean of the independent variable for the prior and sample observations combined. Then,

$$\alpha_0 \sim t\left[m + n - 2, a^{**} + b(x_0 - \bar{x}_{mn}), \left(\frac{1}{m + n} + \frac{(x_0 - \bar{x}_{mn})^2}{A}\right)D\right]$$

where α^{**}, b, A, and D are as defined in Table 7.2. Substituting for these terms we obtain

$$\alpha_0 \sim t(74, 10.18, 40.93)$$

The average decrease here of 10.18 millimeters for individuals participating in the full thirty sessions is larger than the value of 8.64 obtained in the first study, and posterior belief about α at x = 30 is considerably more precise than it was following the first study.

* * * * * *

One could carry out the analysis using a non-data-based prior distribution also. Procedures for eliciting a prior distribution for α^*, β and ϕ are included in CADA. Few people are likely to have very specific beliefs about the regression parameters themselves. The prior distribution for these parameters is obtained indirectly. The user is asked to specify percentiles of distributions for the dependent variable at given values of the independent variable. The judgments required of the user are similar to those described in Chapter 6. Here also, the flexibility of the conjugate prior is enhanced by allowing the number of hypothetical prior observations on ϕ to differ from the number on α^* and β. The more general formulas required for this case are discussed in Novick and Jackson (1974).

COMPARISON OF GROUP MEANS

Regression analysis can also be used in comparing group means. Suppose that we wish to compare two groups in terms of their means on some outcome variable Y. In order to treat this problem using the regression model, we introduce a dummy variable indicating the group to which an individual belongs. An individual is assigned a dummy value of $x = 0$ if the individual is a member of Group 1 and a dummy value of $x = 1$ if the individual is a member of Group 2. Thus for each individual there is a value x, indicating group membership, and a value y, expressing the score on the outcome variable.

Consider again Example 6-4, in which two methods of instruction in mathematics are compared. Group 1 is a control group receiving a standard method of instruction, and Group 2 is an experimental group receiving a new method of instruction. After completion of a unit of material, a 100-point test is administered to assess mastery of the subject. The following results are obtained for the two groups:

$$n_1 = 22 \qquad \bar{y}_1 = 62.27 \qquad S_1^2 = 2809.18$$

$$n_2 = 24 \qquad \bar{y}_2 = 76.22 \qquad S_2^2 = 2888.18$$

where the mean test scores for Groups 1 and 2 are now designated $\bar{y}_1$ and $\bar{y}_2$, respectively.

In order to carry out the regression analysis, we need the sufficient statistics $n, \bar{y}, S_x^2, S_{xy},$ and S^2, along with $\bar{x}$. These could be obtained from the raw data. However, they can also be calculated from the statistics just provided. With some algebraic manipulation, one can show

$$n = n_1 + n_2 \qquad\qquad \bar{y} = \frac{n_1\bar{y}_1 + n_2\bar{y}_2}{n_1 + n_2}$$

$$S_x^2 = (n_1^{-1} + n_2^{-1})^{-1} \qquad S_{xy} = (n_1^{-1} + n_2^{-1})^{-1} (\bar{y}_2 - \bar{y}_1)$$

$$S^2 = S_1^2 + S_2^2 \qquad\qquad \bar{x} \doteq \frac{n_2}{n_1 + n_2}$$

Substituting the appropriate values in these equations, we have $n = 46, \bar{y} = 69.55, S_x^2 = 11.48, S_{xy} = 160.12, S^2 = 5697.36,$ and $\bar{x} = .52$. Using these values in analysis based on noninformative priors, we obtain the marginal posterior distributions

$$\phi \sim \chi^{-2}(44, 5697.36)$$

$$\alpha^* \sim t(44, 69.55, 123.86)$$

and

$$\beta \sim t(44, 13.95, 496.36)$$

We also determine the marginal posterior distribution of the intercept at x = 0, which is $\alpha_0 \sim t(44, 62.27, 258.97)$ and at x = 1, which is $\alpha_0 \sim t(44, 76.22, 237.39)$.

In Example 6-4, the following posterior marginal distributions were obtained:

$$\phi^{1/2} \sim \chi^{-1}(44, 75.48)$$

$$\theta_1 \sim t(44, 62.27, 258.97)$$

$$\theta_2 \sim t(44, 76.22, 237.39)$$

$$\delta \sim t(44, 13.95, 496.36)$$

Because the standard deviation $\phi^{1/2}$ has a χ^{-1} distribution, the variance ϕ is distributed $\chi^{-2}(44, 5697.36)$. This is the same result as was obtained in the regression analysis. We can also see that the posterior distribution of θ_1 is the same as the distribution of the intercept at x = 0. This makes sense; when x = 0 we are concerned with inference for those individuals in Group 1. Likewise, we see that the posterior distribution of θ_2 is the same as the distribution of the intercept at x = 1. Finally, the distribution of $\delta = \theta_2 - \theta_1$ is the same as the distribution of β. That is, β expresses the difference between the means in the control and experimental group. The slope parameter β represents the amount of change in the dependent variable with a change of one unit in the independent variable. Because the change from x = 0 to x = 1 is a change of one unit in the independent variable (group membership), β expresses the amount of change in test score means when going from control to treatment group.

A NOTE ON BAYESIAN INFERENCE
FOR CORRELATION COEFFICIENTS

The discussion up to this point has not involved any assumptions about the distribution of X. In Bayesian regression, the values of X may be preselected or they may be jointly randomly sampled along with the values of Y. When X is randomly sampled with Y, the correlation coefficient may be of interest. We assume here that the model for the (x, y) pairs is bivariate normal. This model has five parameters: θ_X, ϕ_X, θ_Y, ϕ_Y, and ρ. The first four parameters are the means and variances of X and Y, respectively, and ρ is the correlation coefficient expressing the degree of association between X and Y.

For a noninformative prior, we will take $p(\theta_1)$, $p(\theta_2)$, $p(\ln \theta_1)$, and $p(\ln \theta_2)$ to be uniform and independent. Furthermore, we will take the prior distribution on the hyperbolic arctangent of ρ, $p(\tanh_\rho^{-1})$ to be uniform and independent of the other priors. The hyperbolic arctangent is a hyperbolic function; the justification for using a prior of this form is discussed by Lindley (1965b). I will point out below how the implications of this prior are consistent with material previously discussed.

The sample sufficient statistics are n, $\bar{x}$, S_x^2, $\bar{y}$, S_y^2, and S_{xy}, as defined earlier. Given the prior distribution, and these sample statistics, one can make inferences about each of the parameters of the bivariate normal model as well as the regression parameters for the regression of Y on x, or the regression of X on y. We will focus here on inference for ρ.

The posterior marginal distribution of ρ does not take the form of any of the commonly used families of distributions. Inference is carried out using ζ, the hyperbolic arctangent of ρ, which is approximately normally distributed. Let $r = S_{xy}/(S_x^2 S_y^2)^{1/2}$ and let z represent the hyperbolic arctangent of r. Then the posterior distribution of ζ is approximately normal with mean z and variance n^{-1}. The transformation r to z is simply the Fisher r to z transformation that is discussed in most classical treatments of inference for correlation coefficients. Here we use this same transformation to obtain ζ from ρ. Tables for this transformation are included in statistics texts, and this hyperbolic function is available on many pocket calculators. Using the normal distribution, we can make probability statements about ζ, and then transform ζ back into ρ in order to express our inferences in terms of ρ.

Note that, taking the prior $p(\tanh_\rho^{-1})$ to be uniform means that the prior distribution for the transformed variable ζ is uniform. Because ζ is normally distributed with the variance known to be n^{-1} (for fixed n), the appropriate noninformative prior is one that is uniform. Thus the choice of this particular prior for ρ leads to an analysis for the normally distributed parameter ζ that corresponds exactly with that discussed in Chapter 5.

As an example, suppose we obtain r = .73, with n = 50. Transforming r, we have z = .93. The distribution of ζ is therefore approximately normal with a mean of .93 and a variance of $(50)^{-1}$; that is, $\zeta \sim N(.93, .02)$. 95% of the area of a normal distribution lies between $\pm$ 1.96 standard deviations from the mean. The standard deviation here is $(.02)^{\frac{1}{2}}$, thus

$$P[.93 + 1.96\ (.02)^{\frac{1}{2}} \geq \zeta \geq .93 - 1.96\ (.02)^{\frac{1}{2}}] = .95$$

which reduces to $P(1.21 \geq \zeta \geq 0.65) = .95$. Transforming back to ρ by taking the hyperbolic tangent of the endpoints, we obtain $P(.84 \geq \rho \geq .57) = .95$. This is a 95% credible interval for ρ. With vague prior information, the posterior probability that ρ lies in this interval is .95.

Novick and Jackson (1974) discuss methods of incorporating prior information. Gokhale and Press (1982) discuss methods of assessing priors.

MULTIPLE LINEAR REGRESSION

The ideas discussed above extend to the topic of multiple regression. In multiple regression problems, we are interested in the functional relationship between the dependent variable and two or more independent variables. The applications of simple regression are somewhat limited, but multiple regression provides a general framework for the analysis of a wide variety of research designs. We discuss only the case of analysis with noninformative priors; analysis with informative priors is discussed in the readings listed at the end of this chapter.

THE MODEL AND THE LIKELIHOOD FUNCTION

Suppose that we have observations on a dependent variable Y and a set of k independent variables $X_1, X_2, \ldots, X_k$ for each of n cases. It is assumed that the observations are linearly related in the following manner: $y_1 = \alpha + \beta_1 x_{11} + \beta_2 x_{21} + \ldots + \beta_k x_{ki} + e_i, i = 1, 2, \ldots, n$, where y_i is the ith observation on the dependent variable, x_{ji} is the ith observation on an independent variable x_j, e_i is the ith unobserved value of the random error, and $\alpha, \beta_1, \beta_2, \ldots \beta_k$ are regression parameters. As before, the y values are assumed to be randomly sampled, while the x_j values may be either randomly sampled along with the y values or chosen by the investigator. The e's are assumed to be independent and normally distributed with a mean of zero and common variance ϕ. Thus the conditional distribution of Y given $x_1, x_2, \ldots x_k$ is normal with mean $\alpha + \beta_1 x_1 + \beta_2 x_2 + \ldots + \beta_k x_k$, and variance ϕ. The density function for a single observation on Y conditional upon the parameters and the observations is the normal function

$$p(y \mid x_1, x_2, \ldots, x_k, \alpha, \beta_1, \beta_2, \ldots \beta_k, \phi)$$

$$= (2\pi\phi)^{-\frac{1}{2}} \exp \left[\frac{[y - (\alpha + \beta_1 x_1 + \beta_2 x_2 + \ldots + \beta_k x_k)]^2}{-2\phi} \right]$$

The subsequent treatment of multiple regression is most easily presented in terms of matrices and matrix operations. The regression equations for each of the y values are

$$
\begin{aligned}
y_1 &= \alpha + \beta_1 x_{11} + \beta_2 x_{12} + \ldots + \beta_k x_{1k} + e_1 \\
y_2 &= \alpha + \beta_2 x_{21} + \beta_2 x_{22} + \ldots + \beta_k x_{2k} + e_2 \\
&\;\vdots \\
y_n &= \alpha + \beta_1 x_{n1} + \beta_2 x_{n2} + \ldots + \beta_k x_{nk} + e_n.
\end{aligned}
\qquad [7.5]
$$

Now let

$$
y = \begin{bmatrix} y_1 \\ y_2 \\ \vdots \\ y_n \end{bmatrix}
\qquad
X = \begin{bmatrix}
1 & x_{11} & x_{12} & \cdots & x_{1k} \\
1 & x_{21} & x_{22} & \cdots & x_{2k} \\
\vdots & \vdots & \vdots & & \vdots \\
1 & x_{n1} & x_{n2} & \cdots & x_{nk}
\end{bmatrix}
$$

$$
\beta = \begin{bmatrix} \alpha \\ \beta_1 \\ \beta_2 \\ \vdots \\ \beta_k \end{bmatrix}
\qquad
e = \begin{bmatrix} e_1 \\ e_2 \\ \vdots \\ e_n \end{bmatrix}
$$

Then equations 7.5 can be simply written as $\mathbf{y} = \mathbf{X}\,\beta + \mathbf{e}$.

In these terms, the joint density function for n independent observations on Y, given $\mathbf{X}$, β, and ϕ, is

$$p(\mathbf{y}\,|\,\mathbf{X}, \beta, \phi) = 2\pi\phi^{-n/2} \exp\left[\frac{(\mathbf{y} - \mathbf{X}\,\beta)'\ (\mathbf{y} - \mathbf{X}\,\beta)}{-2\phi}\right]$$

With some manipulation the likelihood function can be written

$$\ell(\beta, \phi, \mathbf{y}, \mathbf{X}) \propto \phi^{-n/2} \exp\left[\frac{S^2 + (\beta - \mathbf{b})'\ \mathbf{X}'\mathbf{X}\ (\beta - \mathbf{b})}{-2\phi}\right] \qquad [7.6]$$

where

$$\mathbf{b} = \begin{bmatrix} a \\ b_1 \\ b_2 \\ \vdots \\ b_k \end{bmatrix} = (\mathbf{X}'\mathbf{X})^{-1}\ \mathbf{X}'\mathbf{y}$$

and

$$S^2 = (\mathbf{y} - \mathbf{X}\,\mathbf{b})'\ (\mathbf{y} - \mathbf{X}\,\mathbf{b})$$

The sufficient statistics here are n, $\mathbf{b}$, $\mathbf{X}'\mathbf{X}$, and S^2.

Note that we are not transforming the intercept as we did in the presentation of simple regression. Although we could make this transformation, it does not simplify this discussion to the extent that it did in the case of simple regression. A presentation of Bayesian multiple regression involving this transformation can be found in Laughlin (1979).

ANALYSIS WITH A NONINFORMATIVE PRIOR

The noninformative joint prior is the product of independent uniform priors on α, β_1, β_2, . . . , β_k and ln ϕ. Thus

$$p(\beta, \phi) \propto \phi^{-1} \qquad [7.7]$$

Multiplying the likelihood 7.6 by the prior 7.7 we obtain the posterior

$$p(\beta, \phi | \mathbf{X}, \mathbf{y}) \propto \phi^{-(n/2)-1} \exp \left[\frac{S^2 + (\beta - \mathbf{b})' \, \mathbf{X}' \mathbf{X} \, (\beta - \mathbf{b})}{-2\phi} \right] \qquad [7.8]$$

The marginal distribution of the variance ϕ, obtained by integrating out β, is $\phi \sim \chi^{-2} (\nu, S^2)$ where $\nu = n - k - 1$.

When ϕ is known the posterior 7.8 reduces to

$$p(\beta | \phi, \mathbf{X}, \mathbf{y}) \propto \exp \left[\frac{(\beta - \mathbf{b})' \, \mathbf{X}' \mathbf{X} \, (\beta - \mathbf{b})}{-2\phi} \right]$$

which is the kernel of a multivariate normal distribution with mean vector $\mathbf{b}' = (a, b_1, b_2, \ldots b_k)$ and covariance matrix $\phi \, (\mathbf{X}'\mathbf{X})^{-1}$. This joint conditional distribution of β is written

$$\beta \sim N \, [\mathbf{b}, \phi \, (\mathbf{X}'\mathbf{X})^{-1}]$$

The marginal distribution of α is

$$\alpha \sim N \, (a, \phi \, c_{11})$$

where c_{11} is the first diagonal element of $(\mathbf{X}'\mathbf{X})^{-1}$, and the marginal distribution of β_i is

$$\beta_i \sim N \, [b_i, \phi \, c_{jj}]$$

where c_{jj} is the $i^{th} + 1$ diagonal element of $(\mathbf{X}'\mathbf{X})^{-1}$.

When ϕ is unknown, the marginal distribution of β, obtained by integrating out ϕ, is a multivariate t distribution:

$$\beta \sim t \, [n - k - 1, \mathbf{b}, S^2 \, (\mathbf{X}'\mathbf{X})^{-1}]$$

The first term is the degrees of freedom, the second term is a mean vector, and the third term determines the scale of the distribution. The marginal distribution of α is

$$\alpha \sim t \, [n - k - 1, a, S^2 \, c_{11}]$$

and the marginal distribution of β_i is

$$\beta_i \sim t\,[n - k - 1,\, b_i,\, S^2\, c_{ii}]$$

Readers familiar with the classical theory of multiple regression will recognize parallels with the Bayesian theory outlined above. The means of the marginal posterior distributions of α and β_i correspond with the classical least squares estimates of α and β_i. In the Bayesian approach the posterior marginal distribution of a regression parameter β_i is such that

$$t = \frac{\beta_i - b_i}{(\nu^{-1}\, S^2\, c_{jj})^{\frac{1}{2}}}$$

has a standard t distribution on $n - k - 1$ degrees of freedom, and this can be used in making inferences about the parameter β_i. In the classical approach the distribution of b_i is such that

$$t = \frac{b_i - \beta_i}{(\nu^{-1}\, S^2\, c_{jj})^{\frac{1}{2}}}$$

has a standard t distribution on $n - k - 1$ degrees of freedom, and this in testing hypothesis concerning β_i. In the classical approach, however, it is b rather than β that is random.

* * * * * *

EXAMPLE 7-3. EVALUATING A NUTRITION EDUCATION PROGRAM

A community hospital offers a weight loss program for obese patients. One component of this program involves nutrition education. A two-evening program focuses on the selection of foods and food quantities that provide adequate nutrition while allowing the individual to reduce caloric intake. The developers of the program are interested in evaluating two different methods of presenting this material. The first method of presentation consists of lecture plus videotape programs. The second method of presentation involves lecture plus actual preparation and tasting of certain dishes. In the initial phase of the evaluation, the effectiveness of the two methods is assessed in terms of posttreatment knowledge of the relevant nutritional information.

Interested persons are randomly assigned to the two treatment conditions, with 20 individuals in each group. At the first session, each participant fills out a questionnaire containing items relating to background and demographic characteristics, plus twelve items concerning nutrition and nutrition terminology. At the end of the second session an eighty-item test of nutrition knowledge is administered. Eighteen members of the group receiving the first method of presentation attended both sessions; sixteen members of the other group attended both sessions.

The program developer wishes to compare the two groups on the basis of the nutrition test scores. This can be done using a regression model, as was discussed above. It is expected here that the posttreatment test scores will be related to nutrition knowledge before the sessions, as measured by the twelve nutrition-related items included on the presession questionnaire. Thus in order to improve the precision of the findings, a regression-model analysis of covariance is carried out using the number of the twelve questionnaire items that were answered correctly as a covariate.

The raw data are shown in Table 7.3. The first column contains the case number. Test score y is shown in column 2. The third column contains x_1, the score on the twelve questionnaire items. The fourth column contains x_2, a dummy variable indicating which treatment was received: Persons in Group 1 (lecture plus videotape) are assigned a value of 0, and persons in Group 2 (lecture plus food preparation) are assigned a value of 1. The fifth column contains the product of x_1 and x_2 to be used in examining the interaction between the covariate x_1 and the treatment x_2.

A major assumption in the analysis of covariance is that there is no interaction between treatment and covariates. One would need to determine that this is the case before interpreting the results in analysis of covariance terms. At this point, we simply note that an analysis of these data indicated a lack of interaction, that is, $\beta_3 = 0$ and proceed with the discussion of the main results. We will return to the analysis of the interaction term following this discussion.

TABLE 7.3
Nutrition Knowledge Data

(1) Case Number i	(2) Posttest Score y_i	(3) Questionnaire Score x_{i1}	(4) Treatment Group x_{i2}	(5) Product Term $x_{i3} = x_{i1} x_{i2}$
1	76	12	0	0
2	69	7	0	0
3	68	11	0	0
4	64	6	0	0
5	61	9	0	0
6	60	9	0	0
7	58	8	0	0
8	57	7	0	0
9	57	10	0	0
10	56	7	0	0
11	55	12	0	0
12	54	10	0	0
13	52	7	0	0
14	52	8	0	0
15	51	5	0	0
16	49	3	0	0
17	49	9	0	0
18	45	5	0	0
19	78	10	1	10
20	75	8	1	8
21	75	11	1	11
22	73	6	1	6
23	71	7	1	7
24	71	8	1	8
25	71	11	1	11
26	70	7	1	7
27	69	6	1	6
28	69	7	1	7
29	67	9	1	9
30	64	10	1	10
31	63	12	1	12
32	58	4	1	4
33	54	6	1	6
34	51	4	1	4

The regression model here (with $\beta_3 = 0$) is

$$y_i = \alpha + \beta_1 x_{1i} + \beta_2 x_{2i} + e_i$$

For this problem,

$$
X = \begin{bmatrix} 1 & 12 & 0 \\ 1 & 7 & 0 \\ \cdot & \cdot & \cdot \\ \cdot & \cdot & \cdot \\ 1 & 4 & 1 \end{bmatrix} \qquad y = \begin{bmatrix} 76 \\ 69 \\ \cdot \\ \cdot \\ 51 \end{bmatrix}
$$

One can then calculate

$$
X'X = \begin{bmatrix} 1 & 1 & \ldots & 1 \\ 12 & 7 & \ldots & 4 \\ 0 & 0 & \ldots & 1 \end{bmatrix} \begin{bmatrix} 1 & 12 & 0 \\ 1 & 7 & 0 \\ \cdot & \cdot & \cdot \\ \cdot & \cdot & \cdot \\ 1 & 4 & 1 \end{bmatrix} = \begin{bmatrix} 34 & 271 & 16 \\ 271 & 2353 & 126 \\ 16 & 126 & 16 \end{bmatrix}
$$

$$
(X'X)^{-1} = \begin{bmatrix} .39232 & -.04180 & -.06311 \\ -.04180 & .00519 & .00093 \\ -.06311 & .00093 & .11823 \end{bmatrix}
$$

and

$$
(X'y) = \begin{bmatrix} 1 & 1 & \ldots & 1 \\ 12 & 7 & \ldots & 4 \\ 0 & 0 & \ldots & 1 \end{bmatrix} \begin{bmatrix} 76 \\ 69 \\ \cdot \\ \cdot \\ 51 \end{bmatrix} = \begin{bmatrix} 2112 \\ 17124 \\ 1079 \end{bmatrix}
$$

The values of **b** are obtained as follows:

$$
b = (X'X)^{-1} X'y = \begin{bmatrix} .39232 & -.04180 & -.06311 \\ -.04180 & .00519 & .00093 \\ -.06311 & .00093 & .11823 \end{bmatrix} \begin{bmatrix} 2112 \\ 17124 \\ 1079 \end{bmatrix} = \begin{bmatrix} 44.62 \\ 1.59 \\ 10.34 \end{bmatrix}
$$

From the definition of S^2 we can compute

$$
Xb = \begin{bmatrix} 1 & 12 & 0 \\ 1 & 7 & 0 \\ \cdot & \cdot & \cdot \\ \cdot & \cdot & \cdot \\ 1 & 4 & 1 \end{bmatrix} \begin{bmatrix} 44.62 \\ 1.59 \\ 10.34 \end{bmatrix} = \begin{bmatrix} 63.64 \\ 55.72 \\ \cdot \\ \cdot \\ 80.32 \end{bmatrix}
$$

$$(\mathbf{y} - \mathbf{X}\,\mathbf{b}) = \begin{bmatrix} 76 \\ 69 \\ \vdots \\ 51 \end{bmatrix} - \begin{bmatrix} 63.64 \\ 55.72 \\ \vdots \\ 61.32 \end{bmatrix} = \begin{bmatrix} 12.36 \\ 13.28 \\ \vdots \\ 10.32 \end{bmatrix}$$

and

$$S^2 = (\mathbf{y} - \mathbf{X}\,\mathbf{b})'(\mathbf{y} - \mathbf{X}\,\mathbf{b}) = [12.36 \quad 13.28 \ldots -10.32] \begin{bmatrix} 12.36 \\ 13.28 \\ \vdots \\ -10.32 \end{bmatrix} = 1464.13$$

The degrees of freedom ν for this analysis is $34 - 2 - 1 = 31$. These calculations can be easily carried out using CADA.

The posterior marginal distributions of ϕ, α, β_1 and β_2 are shown in section 1 of Table 7.4. The distribution of primary interest here is that of β_2. Here $\beta_2 \sim t[31, 10.34, 1464.13 \,(.11823)]$. This is the posterior difference between the mean test scores in the two groups, with the effects of the covariate removed or partialed out. The probability that β_2 exceeds zero is

$$P(\beta_2 > 0) = P\left\{ t > \frac{0 - 10.34}{[(31)^{-1} \,(1464.13) \,(.11823)]^{\frac{1}{2}}} \right\} = P(t > -4.38)$$

The probability of a standard t variate with 31 degrees of freedom exceeding this value is over .99. The probability that β exceeds any other value of interest could be determined in the same manner. The 95% HDR runs from 5.52 to 15.16.

The effect of the covariate can be seen by comparing these results with those obtained when y is regressed on the group membership variable x_2 only. From Section 2 of Table 7.4, we see that the posterior distribution of the treatment effect is $\beta_2 \sim t[32, 10.05, 1948.22 \,(.11806)]$. The 95% HDR runs from 4.58 to 15.52. Thus the inclusion of the covariate leads to an increase in the mean of the posterior distribution of β_2 from 10.05 to 10.34. In addition, it increased the precision of posterior belief about the size of the treatment effect; the 95% HDR, for example, was reduced from the interval (4.58, 15.52) to (5.52, 15.16).

TABLE 7.4
Marginal Posterior Distributions of Regression Parameters

(1) Regression of y on x_1 and x_2

$\phi \sim \chi^{-2}$ (31, 1464.13)

$a \sim t$ [31, 44.62, 1464.13 (.39232)]

$\beta_1 \sim t$ [31, 1.59, 1464.13 (.00519)]

$\beta_2 \sim t$ [31, 10.34, 1464.13 (.11823)]

(2) Regression of y on x_2.[a]

$\phi \sim \chi^{-2}$ (32, 1948.22)

$a \sim t$ [32, 57.39, 1948.22 (.05556)]

$\beta_2 \sim t$ [32, 10.05, 1948.22 (.11806)]

(3) Regression of y on x_1, x_2, and x_3 ($= x_1 x_2$).

$\phi \sim \chi^{-2}$ (30, 1463.36)

$a \sim t$ [30, 44.27, 1463.36 (.68592)]

$\beta_1 \sim t$ [30, 1.63, 1463.36 (.00972)]

$\beta_2 \sim t$ [30, 11.067, 1463.36 (1.43940)]

$\beta_3 \sim t$ [30, −0.09, 1463.36 (.02086)]

a. Although there is only one independent variable in the regression equation, the regression coefficient is labeled β_2 for purposes of comparing it with β_2 in regressions (1) and (3) in this table.

Now consider the product interaction term $\beta_3 x_3$. Results obtained when this term is included in the regression are shown in Section 3 of Table 7.4. It can be seen that the mean of the posterior marginal distribution is − .09, which is very close to zero. One possible way to reach a conclusion about whether or not the interaction term is zero is to conduct a Bayesian significance test. One could construct a 95% HDR for β_3 and reject the hypothesis that $\beta_3 = 0$, if the value zero lies outside the boundries of the 95% HDR. In this example, the value zero is well within the 95% HDR (−2.50, 2.32), and the hypothesis that β_3 is zero would not be rejected. Because of the coincidence of the posterior distribution of β_3 and the sampling distribution of b_3, the end points of the 95% HDR coincide with the cutoff points in a classical two-sided t test of the null hypothesis $\beta_3 = 0$, and this null hypothesis would not be rejected in the classical test either. As was discussed in Chapter 5, the Bayesian significance test does not provide a posterior probability that $\beta_3 = 0$. In some ways, it is more in the spirit of classical significance testing than Bayesian comparison of hypotheses. The following section discusses an approach that focuses on the relative posterior probabilities of hypotheses concerning regression models and the regression coefficients.

COMPARING REGRESSION MODELS

In previous discussion of the results of the regression analysis, we have focused on the posterior marginal distribution of the regression coefficient representing the treatment effect. From this posterior, we were able to compare hypotheses about the magnitude of the treatment effect in terms of the posterior probabilities of these hypotheses. For some purposes, the investigator may want to frame certain questions in terms of hypotheses concerning entire regression models. We will consider the case where the investigator wishes to compare two models, one of which is nested in the other, in order to reach some conclusion about which variables should be included in the regression equation. We show how Bayesian methods can be used in examining the interaction term in Example 7-3. In addition, we show how the Bayesian results relate to classical results that are based on consideration of the squared multiple correlation coefficient R^2.

Suppose that the k independent variables can be divided into two groups of k_1 and k_2 variables, where $k_1 + k_2 = k$. Suppose further that we are interested in whether or not to include the set of k_2 variables in the model. Let X_0 and β_0 represent the X matrix and β vector, respectively, for the model including only the k_1 independent variables. Then the question can be framed in terms of comparison of the two hypotheses $H_0: y = X_0 \beta_0 + e$ and $H_1: y = X \beta + e$ where the unsubscripted X and β in H_1 pertain to the model with all k variables. H_0 and H_1 could also be expressed in terms of whether or not the k_2 betas equal 0.

The classical test of these hypotheses can be carried out using the squared multiple correlation coefficient defined as

$$R^2 = \frac{Sy^2 - S^2}{Sy^2} = 1 - \frac{S^2}{Sy^2}$$

The coefficient R^2 expresses the proportion of the variation in the dependent variable that is accounted for by the regression model, and is interpreted as expressing the goodness or completeness of the model in explaining the variation in the dependent variable.

Let R_0^2 and R_1^2 represent the values of this coefficient with the models specified in H_0 and H_1, respectively. Then the statistic

$$F = \left(\frac{R_1^2 - R_0^2}{1 - R_1^2} \right) \left(\frac{n - k - 1}{k - k_1} \right) \tag{7.9}$$

is distributed according to an F distribution with $\nu_1 = k - k_1$ and $\nu_2 = n - k - 1$ degrees of freedom. If the value of the F statistic exceeds the 5% critical value (say) for that F distribution, H_0 is rejected at the 5% level. This test can be thought of as indicating whether the increase in R^2, resulting from including the additional k_2 independent variables, is large enough to be judged statistically significant.

The Bayesian approach here, as in other hypothesis-testing problems, involves computing and comparing the posterior probabilities of the hypotheses. Zellner and Siow (1980: 594) derive the posterior odds ratio of the hypotheses under consideration for the case where prior information is vague and the two hypotheses are considered equally likely a priori. For purposes of comparison with the classical approach, they show how these odds can be expressed as a function of R_0^2 and R_1^2. Specifically with $\Omega' = 1$ the posterior odds ratio in favor of H_0 is shown to be

$$\Omega'' = \left(\frac{\pi^{\frac{1}{2}}}{\Gamma[(k_2 + 1)/2]} \right) \left(\frac{\nu_2}{2} \right)^{k_2/2} \left(\frac{1 - R_1^2}{1 - R_0^2} \right)^{(\nu_2 - 1)/2} \qquad [7.10]$$

where the symbol π denotes the constant 3.14 and Γ indicates the value of the gamma function for the term in brackets.[2]

Consider now the issue of whether the coefficient for the interaction term in example 7-3 is zero. The hypotheses are H_0: $\beta_3 = 0$ and H_1: $\beta_3 \neq 0$. We compute $R_0^2 = .47783$ and $R_1^2 = .47798$. We have $k_2 = 1$ and $\nu_2 = 30$. From a table of values of the gamma function we find $\Gamma[(k_2 + 1)/2] = \Gamma(1) = 1.00$. Entering these values into equation 7.10 we find that the posterior odds in favor of H_0 are

$$\Omega'' = \left(\frac{(3.14)^{\frac{1}{2}}}{1} \right) \left(\frac{30}{2} \right)^{\frac{1}{2}} \left(\frac{1 - .47798}{1 - .47783} \right)^{(30-1)/2} = 6.8236$$

The odds in favor of $\beta_3 = 0$ are close to 7 to 1. The posterior probability of H_0 can be calculated from the odds and is found to be $6.836/(1 + 6.836) = .872$.

The classical test is carried out by entering the appropriate values in equation 7.9. We have

$$F = \left(\frac{.47798 - .47783}{1 - .47798}\right)\left(\frac{30}{1}\right) = .00862$$

The probability of obtaining a statistic of this size or larger if H_0 were true is, according to an F distribution with 1 and 30 degrees of freedom, less then .93, that is, $p < .93$. H_0 would not be rejected at any conventional level.

In this problem both Bayesian and classical procedures lead to the same conclusion, although the posterior probability of H_1 and the classical p-value do not coincide. One problem with the classical test is that for fixed α, any increase in R^2 can be found to be significant with a large enough sample size. We can see from equation 7.9 that F increases as a function of n, and any desired value of F could be obtained with a large enough sample. In the Bayesian analysis, on the other hand, if n increases with other factors held constant, the odds in favor of H_0 increase. This is the same issue that was examined in the discussion of testing point null hypotheses in Chapter 5. In the next chapter we will return to this.

FURTHER READING

Simple regression and correlation with noninformative priors are discussed in Lindley (1965b), Phillips (1973), and Schmitt (1969). Novick and Jackson (1974) discuss these topics using both noninformative and informative priors. Bayesian treatments of multiple regression are found in Broemling (1985), Leamer (1978), Pratt et al. (1965), Raiffa and Schlaifer (1961), and Zellner (1971a). Winkler (1977) and Kadane et al. (1980) discuss assessment of priors.

NOTES

1. The means $\bar{x}$ and $\bar{w}$ are not sufficient statistics. However, they do appear in the calculation of sufficient statistics in the examples.

2. The likelihood ratio is indeterminate with the terms in β uniformly distributed a priori. Zellner and Siow (1980) impose additional restrictions on $p(\beta|\phi)$ to obtain this result. Leamer (1978: ch. 4) discusses other possible restrictions. The solution of Zellner and Siow is of interest here because it permits a comparison with classical inferences using R^2 statistics.

8

Inference for
Binomial Proportions

In this chapter we consider methods appropriate for the analysis of dichotomous data. In evaluations of medical treatments, for example, the outcome measure may be whether the patient was cured or was not cured; or in evaluations of educational methods, the measure of interest may be whether a student passed or failed a course. Our interest is not in particular individuals but rather in groups or populations from which the individuals are sampled; and the focus of the discussion will be on making inferences about the proportion of a population that would be cured, would pass, and so on. We first consider methods for making inferences about a proportion. We then examine some implications of the Bayesian approach for methods of sampling.

INFERENCE FOR A SINGLE PROPORTION

THE MODEL AND THE LIKELIHOOD FUNCTION

The data-generating process here is known as a Bernoulli process, and the two outcomes are often labeled "success" and "failure." The Bernoulli process has one parameter π, which is the probability of a success occurring. When making a number of observations from a Bernoulli process, we will assume that the probability of a success π is constant, or stationary over the observations, and that the observations are conditionally independent given π.

Suppose that we obtain a sample of n observations. We begin by considering the probability that x of these observations will be successes. According to the counting rule for combinations, there are

$$\frac{n!}{x!\,(n-x)!}$$

different ways that x successes could occur in the sample of n observations; this is sometimes written as $_nC_x$ or as $\binom{n}{x}$. Because we are assuming that the observations are conditionally independent, the probability of any particular sequence of x successes and n – x failures is the product of the probabilities of the individual successes and failures: $\pi^x(1-\pi)^{n-x}$. Therefore, with $_nC_x$ different sequences of x successes and n – x failures, the probability of x successes in a sample of n observations given π is

$$p(x\,|\,\pi, n) = \frac{n!}{x!\,(n-x)!}\ \pi^x(1-\pi)^{n-x} \qquad [8.1]$$

The probability distribution over the possible values of x is known as the binomial distribution with parameters π and n.

The likelihood function is written

$$\ell(\pi\,|\,x, n) \propto \pi^x(1-\pi)^{n-x}$$

It is not necessary to retain the combinatorial term in this expression because the relationship is one of proportionality and $_nC_x$ is a constant for given x and n. We see that the observations enter the likelihood only through x, the number of successes, and n, the number of observations. These are sufficient statistics for this model.

PRIOR AND POSTERIOR DISTRIBUTIONS

The natural conjugate family for the binomial model is the family of beta distributions. A beta distribution for the parameter π will be written as

$$p(\pi) = \frac{1}{B(p, q)} \pi^{p-1}(1 - \pi)^{q-1}$$

where $p > 0$ and $q > 0$. The quantities p and q are parameters, in this case hyperparameters, of the beta distribution. Some writers treat $p - 1$ and $q - 1$ as parameters, although this is not the standard notation; the reader should be very careful to note just which quantities are considered as parameters when reading different discussions involving the beta distribution. The term in the denominator is known as the beta function and is defined as

$$B(p, q) = \frac{\Gamma(p)\,\Gamma(q)}{\Gamma(p + q)}$$

where the gamma function of some number x is represented $\Gamma(x)$. When the argument x is an integer, $\Gamma(x) = (x - 1)!$. Thus when p and q are integers

$$B(p, q) = \frac{(p - 1)!\,(q - 1)!}{(p + q - 1)!}$$

The value of the gamma function for a number that is not an integer can be obtained from tables in mathematical handbooks; however, we will not need to calculate $B(p, q)$ here. $B(p, q)$ functions here as a standardizing constant that assures the distribution integrates to unity. The beta distribution with parameters p and q will be denoted as $\beta(p, q)$; again, p and q are the parameters, but the exponents in the beta density function are $p - 1$ and $q - 1$.

The beta distribution takes on a wide variety of forms with different values of p and q. Some representative distributions are shown in Figure

8.1. When p = 1 and q = 1, the distribution is uniform. As p and q both increase, the distribution becomes more bell-shaped in appearance. Note also that when p = q, the distribution is symmetric about $\pi = .50$. When p and q are greater then zero but less than one, the distribution assumes a U-shape. When p and q are both zero, the distribution is not defined.

In describing beta distributions in the following discussion we will refer to the following features:

$$\text{mean} = \frac{p}{p+q} \qquad \text{mode} = \frac{p-1}{p+q-2}$$

and

$$\text{variance} = \frac{\left(\dfrac{p}{p+q}\right)\left(\dfrac{q}{p+q}\right)}{p+q+1}$$

Suppose that prior information about π can be expressed in the form of a beta distribution with hyperparameters a and b:

$$p(\pi) \propto \pi^{(a-1)}(1-\pi)^{(b-1)}$$

Suppose also that the data constitute a sample of n observations from a Bernoulli process with parameter π, in which x successes occur. The posterior is proportional to the product of the prior and the likelihood

$$p(\pi\,|\,x, n) \propto \pi^{(a-1)}(1-\pi)^{(b-1)}\pi^{x}(1-\pi)^{n-x}$$

When rearranged, this is

$$p(\pi\,|\,x, n) \propto \pi^{(a+x-1)}(1-\pi)^{(b+n-x-1)} \tag{8.2}$$

The resulting product can be recognized as the kernel of a $\beta(a + x, b + n - x)$ distribution. The posterior is therefore a β distribution, and the values of p and q are simply sums of the respective prior hyperparameters and exponents of the likelihood.

The family of beta distributions satisfies the criteria for natural conjugate families that were discussed previously. The beta distribution is well-defined and is mathematically tractable. As can be seen in Figure

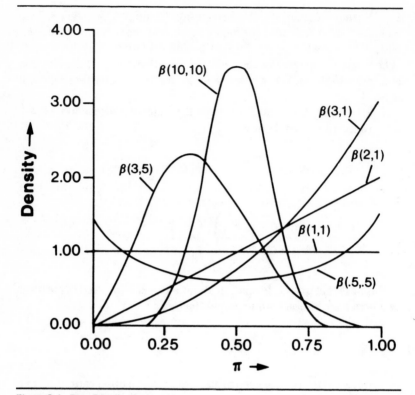

Figure 8.1   Beta Distributions

8.1, it is rich in the shapes that it can assume in representing beliefs about π. Finally, the parameters have a straightforward interpretation: The way that the parameters of the prior combine with x and n – x in expression 8.2 suggests that the prior can be viewed in terms of an equivalent prior sample of a + b observations of which a are successes.

ANALYSIS WITH A NONINFORMATIVE PRIOR

If we think of the prior in terms of an equivalent prior sample of a + b observations with successes, then the improper β (0, 0) prior expresses the notion of lack of prior information. However, if we think of a uniform prior as representing lack of information about π, then the prior should be a β (1, 1) distribution. With this uniform prior, the posterior is proportional to the likelihood; but interpreted as an equivalent prior sample, it is equivalent to having two prior observations, one of which was a success. Arguments based on other considerations lead to a β (.5, .5) prior. A case can be made for any of these priors. We will adopt the prior with a = b = 0 as a noninformative prior because of its ease of interpretation. It is important to keep in mind here that a noninformative prior is not selected because it is the exact representation of ignorance. Rather, it is a representation of information that is imprecise relative to the data. As such, it is a useful approximation that enables us to get started with the analysis. Arguments can be made for different approximations; yet, with even modest-sized samples, any of the priors just discussed will lead to virtually identical results. With very small samples the different priors can lead to different results; however, in such instances it is inappropriate to assume that prior information can be represented as being imprecise in relation to the data, and the uncritical use of any of these noninformative priors in such circumstances can be questioned. If there is little sample data, then one needs to give serious consideration to what prior information one has—choosing some noninformative prior simply because it is convenient will be unsatisfactory. Inference for π using a noninformative β (0, 0) prior is shown in Table 8.1.

* * * * * *

EXAMPLE 8-1. AN ANALYSIS USING A NONINFORMATIVE PRIOR: EVALUATING AN OUTPATIENT PSYCHIATRIC TREATMENT PROGRAM

An outpatient clinic in a community mental health center receives referrals from a state psychiatric facility for all patients discharged to the catchment area for the center following psychiatric hospitalization. An objective of the outpatient program is to enable the patient to function outside of the hospital and thus reduce the need for repeated hospitalization. The director of the clinic is interested in determining the recidivism rate for the referral cases and decides to examine the proportion of patients rehospitalized within one year of their first clinic visit. All referral cases are seen at least once.

A chart review is conducted with a random sample of 20 cases first seen at least one year ago. Information in the chart and provided by the state facility is used to determine whether or not rehospitalization took place. It is found that 7 of the 20 patients were rehospitalized within one year of their first outpatient visit; thus the sample sufficient statistics are x = 7 and n = 20.

TABLE 8.1
Inference for a Binomial Proportion with a
Noninformative Prior Distribution

Prior Distribution:
$\pi \sim \beta$ (0, 0)

Sample Sufficient Statistics:
n, x

Posterior Distribution:
$\pi \sim \beta$ (x, n − x)

Analyzing these data with a noninformative beta prior with a and b equal to zero yields the following posterior distribution:

$$p(\pi|x, n) \propto \pi^{(0+7-1)}(1 - \pi)^{(0+20-7-1)}$$

which we write $\pi \sim \beta$ (7, 13). Again, keep in mind that 7 and 13 are the parameters of the beta distribution, while the exponents in the kernel are 7 − 1 = 6 and 13 − 1 = 12. The mean of the posterior distribution of π is 7/(7 + 13) = .35, and the mode is (7 − 1)/(7 + 13 − 2) = .33, which is not too different from the mean. The variance is

$$\frac{\left(\dfrac{7}{7 + 13}\right)\left(\dfrac{13}{7 + 13}\right)}{7 + 13 + 1} = .01$$

The standard deviation is $(.01)^{\frac{1}{2}}$ or .10. A graph of the posterior distribution of π is shown in Figure 8.2.

The 50% HDR obtained using CADA runs from .26 to .41. These are the limits of the shortest interval to contain 50% of the area of this posterior distribution of π. Exact classical confidence limits for π are not available because the variance of the binomial distribution depends upon the value of π, which is unknown. A classical large-sample approximation is discussed in a subsequent section.

The statewide recidivism rate for the type of patients being referred has been reported as being around 50%. Using CADA, the probability that π is less than .50 is determined to be .92. The program director is, therefore, reasonably confident that the recidivism rate for patients treated in that clinic is lower than the statewide rate.

```
                     BETA DISTRIBUTION
--------------------------------------------------------------------
     ALPHA =   7.00          BETA =  13.00
     MEAN = 0.350      ST. DEV. = 0.104
     MODE = 0.333      SKEWNESS = 0.160
--------------------------------------------------------------------

WHEN YOU ARE READY FOR THE GRAPH TO BE DISPLAYED      TYPE '1'?  1

OPTION 7:    GRAPH OF A BETA DENSITY FUNCTION

   3.72!                    *  *
       !                 *
       !                       *
       !
       !
       !              *            *
       !
       !
       !           *                 *
       !
       !
       !        *                        *
       !
       !      *
   0.00!*  *  *                             *  *  *  *  *  *  *  *  *  *
       !----------+----------+----------+----------+----------+-
       0.00              0.33                          1.00
                       (MODE)
```

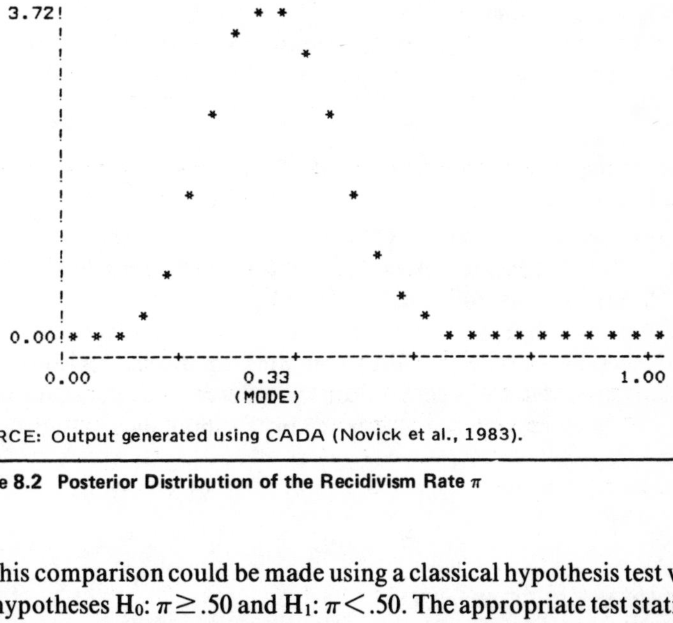

SOURCE: Output generated using CADA (Novick et al., 1983).

Figure 8.2 Posterior Distribution of the Recidivism Rate π

This comparison could be made using a classical hypothesis test with the hypotheses H_0: $\pi \geq .50$ and H_1: $\pi < .50$. The appropriate test statistic is x and the sampling distribution of x is a binomial distribution as in equation 8.1. The probability that seven or less successes would be observed, if the true proportion were .50, is the probability that x = 7, plus the probability that x = 6, and so on down to x = 0. That is,

$$p(x \leq 7 | \pi = .50, \ n = 20) = \sum_{i=1}^{7} \binom{n}{x_i} \pi^{x_i} (1 - \pi)^{(n - x_i)}$$

The p-value here turns out to be .13. This does not coincide with the Bayesian posterior probability that $\pi \geq .50$, which is 1.00 – .92 = 08. There will be no necessary correspondence here between classical p-values and Bayesian posterior probabilities. In the normal theory problems that we considered earlier, the sampling distribution of the sample mean, for example, was in the same general form as the posterior distribution of the population mean, and this led to a numerical coincidence in the classical and Bayesian results. In making inferences about proportion, we find that the sampling distribution of x is a discrete binomial distribution whereas the posterior distribution of π is a continuous beta distribution and there is no reason to expect numerically identical results.

Suppose that the investigator were to analyze the data using a uniform beta prior with a = 1 and b = 1. The posterior distribution is

$$p(\pi|x, n) \propto \pi^{(1+7-1)}(1-\pi)^{(1+20-7-1)}$$

and the posterior parameters are p = 8 and q = 14. The mean of the posterior is .36 and the standard deviation is (with rounding) still .10. The 50% HDR runs from .28 to .42, and $P(\pi < .50) = .91$. In comparing these figures with those obtained earlier, it can be seen that, even with a relatively small sample, the different noninformative priors lead to very similar results. The differences are on the order of .01 and .02, and would be unlikely to make much difference in the thinking of the program director.

* * * * * *

A NORMAL APPROXIMATION

Evaluation of beta distributions without access to a computer program such as CADA can require a fairly extensive set of tables, and certain transformations and approximations to the normal distribution are commonly used to simplify the analysis.

One simple approximation is that when p and q are both 10 or larger, the beta distribution of π is such that π is approximately normally distributed with mean and variance equal to those of the beta distribution. This means that

$$z = \frac{\pi - \dfrac{p}{p+q}}{\left[\dfrac{\left(\dfrac{p}{p+q}\right)\left(\dfrac{q}{p+q}\right)}{p+q+1}\right]^{\frac{1}{2}}} \qquad [8.3]$$

is approximately distributed as a standard normal variate, and one can easily make a variety of probability statements about π by referring to tables for the standard normal distribution.

Analyzing a set of data with a noninformative prior, as shown in Table 8.1, we find that the posterior hyperparameters are p = x and q = n – x, and p + q = n. Substituting for p, q, and p + q in equation 8.3, we have

$$z = \frac{\pi - \dfrac{x}{n}}{\left[\dfrac{\left(\dfrac{x}{n}\right)\left(1 - \dfrac{x}{n}\right)}{n+1}\right]^{\frac{1}{2}}} \qquad [8.4]$$

It is interesting to compare this with the normal approximation to the binomial distribution that is often used in classical analyses when n is large and π of moderate value. Using the approximation, the statistic

$$z = \frac{\dfrac{x}{n} - \pi}{\left[\dfrac{\pi(1 - \pi)}{n}\right]^{\frac{1}{2}}} \qquad [8.5]$$

is approximately distributed as a standard normal variate. Ignoring the difference between n and n + 1, which has little impact with large n, there is a certain symmetry in these expressions, specifically in the appearance of the sample proportion x/n and the population proportion π. When n is large and x/n and π are in the mid-region of the 0 to 1 interval, differences between x/n and π on the order of .10 or even .15 in this mid-region will not greatly affect the denominator. The numerators in the two expressions are the same except for a reversal in sign. Therefore, the resulting z values will be similar (although reversed in sign). Thus p-values obtained in classical one-sided tests and posterior probabilities of hypotheses may be numerically similar with large samples. The interpretation of the results is very different, as I have stressed before.

Classical interval estimation using the normal approximation raises an additional issue not encountered in conducting the classical hypothesis test. In the case of the hypothesis test, the sampling distribution of the test statistic z in equation 8.5 is conditional upon some null value of π

TABLE 8.2
Inference for a Binomial Proportion with an
Informative Prior Distribution

Prior Distribution:
$\pi \sim \beta$ (a, b)

Sample Sufficient Statistics:
n, x

Posterior Distribution:
$\pi \sim \beta$ (a + x, b + n − x)

specified in the null hypothesis. The investigator simply enters this value π_0 for π in equation 8.5 and obtains the sampling distribution of z, conditional upon H_0 being true. In the case of classical interval estimation the problem is complicated by the fact that π is unknown and the standard error term involves this unknown quantity. In order to proceed, the sample estimate x/n is entered in the standard error term in place of π. This makes the denominators in equations 8.4 and 8.5 the same, except for the difference between n and n + 1. Consequently, when n is large, classical confidence intervals and Bayesian credible intervals will be almost identical numerically.

ANALYSIS WITH AN INFORMATIVE PRIOR

As we saw earlier, analysis with informative conjugate priors is fairly simple; inference in this case is summarized in Table 8.2. Data-based priors will generally be in the appropriate form such that the posterior parameters can be obtained by simple addition of prior hyperparameters and sample sufficient statistics. Likewise, the expression of non-data-based priors in terms of equivalent prior samples places them in this form. There are various methods for quantifying non-data-based prior information. The CADA procedures for fitting prior information about π with a beta prior are similar to those for fitting prior information about normal means and standard deviations with normal and inverse chi distributions. The user is asked to specify percentiles, and is then presented with several alternative fits to these percentiles. The equivalent prior sample size and the mean of the distribution can be adjusted until the user is satisfied that the features of the prior correspond with his or her beliefs about π. Actually, assessment of prior beliefs for the binomial model is less complicated than it is for the normal model with unknown variance because the user need only assess a prior distribution for one parameter π rather than a joint distribution of two parameters θ and ϕ. Phillips (1973: 126-127) presents an alternative procedure where a user first selects a most likely or modal value of π. The user then refers to a figure displaying a number of beta distributions with a mode at the selected value. These distributions differ in the values of p and q, and in the bounds of the 50% HDRs. The user selects the distribution for which the 50% HDR seems most acceptable. The figures appear as Appendix B in Phillips's text. Pitz (1982) presents a tabular method of quantifying prior information. The user is asked to specify tertiles that divide the range of π into three equally likely areas. Given the two values t_1 and t_2 that divide the beta distribution on the interval $0 - 1$ into three equal portions, the beta distribution is determined. The user simply refers to a table of values for t_1 and t_2 and reads off corresponding values for the parameters of the beta distribution. Pitz uses a parameterization of the beta distribution that is different from that used in this book and the reader should refer to definitions in that article before using the tables.

$$* \quad * \quad * \quad * \quad * \quad *$$

EXAMPLE 8-2. AN ANALYSIS USING AN INFORMATIVE PRIOR: EVALUATING AN OUTPATIENT PSYCHIATRIC TREATMENT PROGRAM

For quality assurance purposes, the program director conducts a reaudit of recidivism rates for referrals, six months after the initial study. Again, a sample of 20 cases is obtained and it is found that 4 of the 20 patients were rehospitalized. Using the posterior from the previous study as a prior distribution, the prior parameters are a = 7 and b = 13. Then, with sufficient statistics x = 4 and n = 20

$$p(\pi | x, n) \propto \pi^{(7+4-1)}(1-\pi)^{(13+20-4-1)}$$

which we write $\pi \sim \beta (11, 29)$.

The mean is now $11/(11 + 29) = .28$ and the mode is $(11 - 1)/(11 + 20 - 2) = .26$. The variance is

$$\frac{\left(\dfrac{11}{11 + 29}\right)\left(\dfrac{29}{11 + 29}\right)}{11 + 29 + 1} = .0049$$

and the standard deviation is .07. The 50% HDR runs from .22 to .31. The accumulation of data has led to more certain belief about π. This is not a necessary result; under certain conditions, new data that are highly discrepant from the prior will decrease certainty about π.

*   *   *   *   *   *

The determination of the posterior here is very simple and the calculations do not require much discussion. One important point that was mentioned earlier is that, from the Bayesian perspective, it makes no difference whether the first sample is analyzed by itself and then used as a prior in the analysis of the second sample, or whether the total set of observations are analyzed as one sample. The posterior obtained in Example 8-2 is the same as the posterior that would be obtained starting with a = 0 and b = 0 and proceeding as if one had a single sample with 11 successes in 40 observations. An even more important aspect of this is that it does not matter whether the two samples were planned at the beginning of the study, or the decision to obtain a second sample was made after seeing the results of the first sample. The posterior distribution will be the same in either case. In the classical approach, the method of sampling does affect the results. We consider this in the following section.

STOPPING RULES

A stopping rule simply specifies when the investigator should stop collecting data. This section considers implications for classical and Bayesian inference of two different stopping rules for sampling from a Bernoulli process, and shows how the Bayesian approach leads to results that are quite different from results obtained in the classical approach.

Binomial Versus Pascal Sampling

Up to this point, we have talked about drawing a sample of n observations and then counting the number of successes x that occurred in that sample. For this binomial sampling, the probability of x successes, as was discussed earlier, is

$$p(x \mid \pi, n) = \binom{n}{x} \pi^x (1 - \pi)^{n-x} \qquad [8.6]$$

An alternative method of sampling that could result in x successes in n observations is that of making observations sequentially until exactly x successes are obtained. This is known as Pascal sampling; the probability of making n observations to obtain x successes is

$$p(n \mid \pi, x) = \binom{n-1}{x-1} \pi^x (1 - \pi)^{n-x} \qquad [8.7]$$

Because this involves sampling until x successes are observed, the last observation will always be a success; therefore, there are $\binom{n-1}{x-1}$ combinations of the preceding successes, and the combinatorial term in equation 8.7 therefore differs from that in equation 8.6. These two methods of sampling are based on different rules concerning when to stop collecting data. In the case of binomial sampling, the stopping rule says to collect data until n observations are obtained; in the case of Pascal sampling, the stopping rule says to collect data until x successes are observed. Thus we have two possible routes to the same observed data.

When making inferences about π in the classical approach, these stopping rules affect the sample space and resulting inference. Suppose that we obtain 2 successes in 10 observations with binomial sampling, and wish to test the hypotheses H_0: $\pi \geq .50$ and H_1: $\pi < .50$. The probability of obtaining this result, or a more extreme result if H_0 is true, is the sum of the probabilities of obtaining 0, 1, and 2 successes when $\pi = .50$ and n = 10. Using equation 8.6, we calculate each of these probabilities. Their sum, which is the p-value, is .055, and H_0 would not be rejected at the 5% level. Now suppose that the 2 successes in 10 observations were obtained through Pascal sampling. The probability of obtaining this result or a more extreme result if H_0 were true is the sum of the probabilities of making 10, 11, 12, and so on, observations before 2 successes were observed. For purposes of calculation, this is the same as one minus the sum of making 2 through 9 observations before obtaining 2 successes, with $\pi = .50$. Using equation 8.7 this latter sum is found to be .98 and the p-value is $1.00 - .98 = .02$. In this case H_0 would be rejected at the 5% level.

In the Bayesian approach, these different stopping rules have no effect on the results. The likelihood function for π given x and n is proportional to $\pi^x (1 - x)^{n-x}$ in both types of sampling. Consequently, the posteriors will be the same. If these data were analyzed with a noninformative prior $\pi \sim \beta(0, 0)$, for example, the posterior would be π

$\sim\beta(2, 8)$ in both cases, and the posterior probability that $\pi \geq .50$ would be .02 irrespective of which sampling method was used.

Thus, in the classical approach, there are two different long runs as expressed in the different sampling distributions, and the conclusions vary depending upon the route by which the data were obtained. In the Bayesian approach, the likelihood functions given the observed data are the same for both methods of sampling and the conclusions about π are the same.

The difference here hinges on the *likelihood principle*. At the heart of the likelihood principle is the idea that, for any experiment, all the information concerning the parameters of interest is contained in the likelihood function, as determined by the data that were observed. Other features of the experiment that do not affect the likelihood function should not affect the inference. More specifically, the likelihood principle states that any two experiments (concerning the same unknown parameter) with the same likelihood functions should lead to the same inferences. The likelihood principle is implied by Bayes' theorem; however, it is in direct conflict with the classical sampling theory approach where central concepts such as significance, power, and unbiasedness depend upon features of the experiment over and above those expressed in the likelihood.[1]

The likelihood function is conditional upon the data that were observed. It is only the observed data that enter the analysis; other data or results that might have been obtained, but were not, are irrelevant to the analysis. In the classical approach, results that might have been obtained, but were not, are essential to the analysis. In computing p-values, for example, the probability is not only a function of the obtained value of the test statistic if H_0 is true but also a function of the probability of other values of the test statistic that are even less likely under H_0, but that were not obtained. Bayesians argue that once the data are obtained, these other values are irrelevant for inference regarding H_0. In commenting on the fact that results that are unlikely under H_0 and are not observed enter into the p-value and subsequent decision about H_0, Jeffreys (1961: 385) writes, "A hypothesis that may be true may be rejected because it has not predicted observable results that have not occurred. This seems a remarkable procedure."

The Accumulation of Data

Some further consequences of this difference in Bayesian and classical theory can be seen in a problem discussed by Cornfield (1966, 1970).

An investigator using classical methods conducts an experiment hoping to reject H_0 at the .05 level, on the basis of n observations. Analysis of the data reveals that the results are not significant at this level. If they had been significant the investigator would have stopped at that point. However, the investigator feels that H_0 is false and would like to collect some additional data in order to be reasonably certain of rejecting H_0, if it indeed is false. The investigator then obtains an additional m observations. The combined sample of n + m observations permits rejection of H_0 at the .05 level.

Again the route by which the data were obtained is important in the classical analysis. There are two different long runs here depending upon how one views the steps in the collection of the data, and these long runs affect the sampling distributions and subsequent inferences. If the long run is seen as an infinite number of repeated experiments with a sample size of n + m, then .05 is the probability of erroneously rejecting H_0 when it is true. However, if the long run is viewed as an infinite number of repeated 2-stage experiments, the particular results obtained by the investigator would not permit rejection of H_0 at the .05 level. In the first stage, there is a .05 chance of erroneously rejecting H_0 if it is true and a corresponding .95 chance of not rejecting it. The probability of not rejecting H_0 in stage 1 and then erroneously rejecting it with the combined samples in stage 2 approaches $.95 \times .05 = .0475$ as m increases relative to n. The total probability of erroneously rejecting H_0 for the two-stage experiment, then, is the probability of erroneously rejecting at stage 1 or at stage 2, which is $.05 + .0475 = .0975$, assuming that both tests are carried out at the .05 level.

Thus an investigator who took one combined sample of n + m observations could obtain a result permitting rejection of H_0 at the .05 level; yet the investigator who obtained the same result in the 2-stage procedure could not reject H_0 at that level. In fact, no matter what significance level is used in the second-stage testing with the combined sample, there is no result with any amount of data that would ever permit rejection of H_0 at the .05 level in the 2-stage experiment. For an investigator hoping to reject H_0 with $\alpha = .05$, at least, further collection of data would be wasted effort after failure to reject in stage one.

In the Bayesian approach, the likelihood and the posteriors are the same whether the data were collected all at once or in two stages. Consider the two-stage analysis. After obtaining the data in the first stage we have

$$p(\theta \mid x_1) \propto p(\theta)\, \ell(\theta \mid x_1) \qquad [8.8]$$

where x_1 is the data obtained in stage 1. Then, if the investigator collects additional data x_2, conditionally independent of x_1 given θ, in a second stage, we have

$$p(\theta \mid x_1, x_2) \propto p(\theta \mid x_1)\, \ell(\theta \mid x_2) \qquad [8.9]$$

where the posterior from stage 1 is now the prior for stage 2. Substituting the right-hand side of expression 8.8 for $p(\theta \mid x_1)$ in expression 8.9, we obtain

$$p(\theta \mid x_1, x_2) \propto p(\theta)\, \ell(\theta \mid x_1)\, \ell(\theta \mid x_2) \qquad [8.10]$$

On the other hand, if the data were all collected in a single stage then we would have

$$p(\theta \mid x_1, x_2) \propto p(\theta)\, \ell(\theta \mid x_1, x_2) \qquad [8.11]$$

The combined likelihood over the different sets of data is the product of the separate likelihoods (Edwards, 1972: ch. 2), so that in this case

$$\ell(\theta \mid x_1, x_2) \propto \ell(\theta \mid x_1)\, \ell(\theta \mid x_2)$$

Therefore, the posteriors in expression 8.10 and 8.11 for the 2-stage and the single-stage studies, respectively, are the same.

In contrast to the investigator using classical methods, the Bayesian investigator would reach the same conclusion with either method of data collection. The Bayesian investigator could examine the results at any point and make decisions about terminating the study or collecting additional data with a minimum of analytic complexity. The Bayesian approach, therefore, allows a great deal more flexibility than does the classical approach in the collection and analysis of data.

The Persistent Investigator

Persons holding classical views express concern that this argument concerning stopping rules will encourage the "persistent investigator" to capitalize on chance by continuing to collect data until the null hypothesis can be rejected even when it is true. Consider an example where an investigator wishes to reject a point null hypothesis concerning a binomial parameter π, H_0: $\pi = .50$, and accept the alternative hypothesis H_1: $\pi \neq .50$. Suppose that the investigator obtains a sample and determines whether H_0 can be rejected using a classical test; if it cannot be rejected, the investigator obtains another sample, combines it with the first, and repeats this process. If this continues long enough, the chance of rejecting this null hypothesis at any level α, even when $\pi = .50$, approaches one. Thus it would seem that ignoring the stopping rule leads to a highly undesirable consequence.

This problem does not arise in the Bayesian approach. In order to proceed, we will assume that the Bayesian investigator assigns some appreciable prior probability to the point null hypothesis. The comparison of two hypotheses can be carried out using the odds-likelihood ratio form of Bayes' theorem. To simplify the analysis, suppose that the investigator uses a noninformative prior assigning the same probability of .50 to both H_0 and H_1. The prior odds are therefore $.50/.50 = 1.00$. Jeffreys (1961: 256) provides the following expression for the likelihood ratio in making inferences about a binomial proportion:

$$LR = \frac{(n+1)!}{x!(n-x)!} \, \pi_0^x (1 - \pi_0)^{n-x}$$

and provides an approximation not involving factorial terms, for use when x and n – x are large. Given x and n, the posterior odds in favor of H_0 are calculated by multiplying the prior odds by the likelihood ratio. Because the prior odds ratio is 1.00, the posterior odds in this case will be the same as the likelihood ratio. For this example $\pi_0 = .50$.

An interesting comparison between classical and Bayesian inferences as data accumulate can be made by plotting the Bayesian posterior odds in favor of the null hypothesis for results that would be just significant in a classical test at the .05 level, for various sample sizes. This plot for sample sizes from 50 on up to 1000 is shown in Figure 8.3. We see that results that would lead to rejection of H_0 in a classical test, at the same .05 level in samples of different sizes, have very different implications in the Bayesian analysis. Results that lead to a classical rejection of H_0 at the .05 level, with a sample of 50, yield posterior odds less than one and are therefore evidence against H_0 to a Bayesian also. However, as n becomes larger, results that lead to rejection of H_0 in the classical test are, in the Bayesian framework, evidence in favor of H_0. For example,

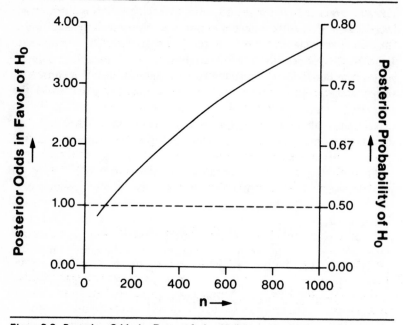

Figure 8.3 Posterior Odds in Favor of the Null Hypothesis When Classical Test
Results Are Just Significant at the .05 Level

with n = 1000, results that would lead the user of classical methods to
reject H_0 would lead the Bayesian to conclude that H_0 is almost four
times more likely to be true than is H_1. As the reader may have recognized, this is just another expression of Lindley's paradox, which was
discussed in Chapter 5.[2]

Implicit in the concern over the persistent investigator is what Cornfield (1966) has labeled the α-postulate: All null hypotheses rejected at
the same α-level have the same amount of evidence against them.
However, as Figure 8.3 shows, a rejection of H_0 with a small sample
means something very different than rejection of H_0 at the same level
with a large sample. From the Bayesian perspective, the α-postulate is
simply not true, and α is an inadequate indicator of evidence against H_0.
Once one acknowledges this, the problem of the persistent investigator
disappears. It is true that the investigator will inevitably be able to reject
H_0 with some fixed α as the data accumulate; however, when properly
interpreted, the results that permit this inevitable rejection may very well
be evidence in favor of H_0 rather than against it. For the Bayesian
investigator, the evidence from the sample is expressed in the likelihood

function rather than in a significance level, and continued collection of data does not guarantee an inevitable accumulation of data against H_0. The difference between the two approaches arises from the fact that the Bayesian analysis is based on the likelihood function whereas the classical analysis is based on the sampling distribution that depends upon other features of the experiment. Writes Savage (1962: 18),

> The likelihood principle . . . affirms that the experimenter's intention to persist does not change the import of his experience. The true moral of the facts about optional stopping is that significance level is not really a good guide to "level of significance" in the sense of "degree of import," for the degree of import does depend on the likelihood alone.

COMPARISON OF PROPORTIONS

Suppose that we want to compare two proportions π_1 and π_2. Inference concerning the difference $\pi_2 - \pi_1$ is complicated because the distribution of the difference between two beta-distributed variables is not of a form that can be referred to commonly tabulated distributions. Consequently, there are a variety of transformations and approximations in use. If we use the normal approximation to the beta distribution that was discussed earlier, and our priors and samples are independent, the posteriors will be independent normally distributed variables with known variances. Comparisons can then be carried out using the methods described in Chapter 5. Other comparison techniques are discussed in Lindley (1965b) and Novick and Jackson (1974). When using computer programs such as CADA or those described in Schlaifer (1971), these methods are not necessary because the probabilities of interest can be computed directly.

FURTHER READING

Introductions to inference for binomial proportions can be found in Good (1965), Lindley (1965b), Novick and Jackson (1974), Phillips (1973), Pratt et al. (1965), Raiffa and Schlaifer (1961), Schmitt (1969), and Winkler (1972). Pitz (1968) compares classical and Bayesian inference for a binomial proportion. In the Bayesian approach, the assumption of independent events is sometimes replaced with a weaker assumption of exchangeable events; Lindley and Phillips (1976) provide an introductory discussion of this concept and its implications in sampling from a Bernoulli process.

NOTES

1. Arguments for the likelihood principle and for inference based on the likelihood function can be made on non-Bayesian grounds also. These ideas can be found in some of Fisher's theories; for example Fisher (1973: ch. 5); indeed, the use of the term "likelihood" in its technical sense is due to Fisher. Birnbaum (1962a, 1962b) provided an argument for the likelihood principle based upon considerations of statistical evidence, apart from those entailed in the Bayesian approach, that stimulated much discussion. Edwards (1972) provides a comprehensive treatment of likelihood inference. Non-Bayesian likelihood inference is an alternative to classical Neyman-Pearson theory and Bayesian theory; however, much of the discussion surrounding the likelihood principle has been highly theoretical in nature, and likelihood inference has yet to be widely applied outside of the Bayesian framework. Some brief reviews of this topic and further references can be found in Barnett (1982: 282-287), Birnbaum (1978), and Joshi (1983). Sprott and Kalbfleisch (1965) discuss the potential use of likelihood inference in psychological research.

2. The binomial distribution is discrete and p-values will vary slightly for results that are just significant at the .05 level. With $n = 50$, the smallest result for which $p < .05$, has a p-value of .032; with $n = 100$ the p-value is .044. With larger samples, the p-values become increasingly closer to .05. These differences do not affect the general conclusion here.

9

Decision Analysis

In this chapter we look at how research findings can be used in decision making. Evaluation research is commonly advocated on the grounds of aiding decision making, and we share that view. Previous chapters considered how belief should be revised on the basis of new data; here we examine how these revised beliefs should be taken into account in decision making. We first examine the classical decision-making procedure of hypothesis testing in some detail and then illustrate the use of Bayesian decision theory. The decision problem is more complex than the inference problem. In addition to uncertainty concerning unknown quantities or states of the world, one must consider alternative actions, their uncertain consequences, and the desirability of the consequences. A thorough discussion of how to carry out decision analysis is beyond the scope of this book. Our goal here is to indicate how evaluation research findings can be coherently combined with other information in decision making. Several references on decision analysis are cited at the end of the chapter for the reader who wishes to pursue the topic further.

Given that evaluation is often seen as a decision aid, it is interesting that relatively little attention has been given to just how evaluation results should enter into decision making. Indeed, the notions of decision making implied in many discussions of evaluation are quite vague. This situation has begun to change, however, with the development of explicit models of evaluative decision making based on multiattribute decision theory, and this line of thinking derives from the general Bayesian decision framework discussed in Chapter 3. The purpose here is to look at how individuals (and, in some instances, groups) can use the results of evaluation in decision making.

It is unfortunate that the term "decision making" typically calls to mind major policy decisions. Such decisions are often the result of complex multiparty interactions and it may seem that no one actually makes decisions. Yet as Keeney and Raiffa (1976) note, decisions, as the term is used here, do not have to be these "grandiose end determinations." Individuals do make decisions such as what course of action to advocate, how to vote on an issue, and when to introduce a proposal, and the theory can be of considerable value on that level. Its use is not limited to the individual level, however. An analysis of a group decision problem may help clarify the issues involved and pinpoint areas of disagreement needing resolution.

Before discussing the details of Bayesian decision theory, it is important to put the discussion into perspective. Decision making is part of the process of solving problems and, in a broad sense, decision making includes identifying a problem, identifying alternative actions or solutions to the problem, evaluating the alternatives, and making a choice. We will be concerned with the use of research findings in evaluating the alternatives; specifically, we will consider how posterior distributions expressing uncertainty about program treatment effects and outcomes can be formally incorporated into the analysis for this purpose. An evaluation report, however, is a complex stimulus made up of many elements, and is much more than a simple statement about treatment effects. It often includes discussion, interpretation, references to other work, and so on. Each of these aspects of the report may be useful in the different phases of decision making. In some cases, the actual measure of treatment effect may have less impact on decision making than some conceptual discussion contained in the report, which influences the decision maker's view of the problem. Indeed, even the color and layout of the report, independent of its content, could trigger some association that aids the decision maker in solving the problem. The point here is that evaluations can affect decision making in a complex variety of ways, and the relevant theories for conceptualizing these myriad effects may lie as much in the psychology of creativity and insight as in statistical decision theory. This is a book on statistics, and I do not propose to treat this entire area of decision making. The problems that will be considered are those in which the decision maker has defined the problem, identified the alternatives, and wishes to examine the implications of the alternatives formally before making a choice.

By focusing on the formal aspects of decision analysis I am not denying the importance of the other aspects of decision analysis. On the contrary, they are critical in bringing the decision problem to the point where formal methods can be of value. Brown et al. (1974: 1) write the following:

> Decision analysis includes four quite dissimilar operations: the construction of problem models; the selection of numerical input for these models; the derivation of model output from inputs; and the interpretation and use of this output as a guide to choice. Only one of these operations—the derivation of model output from inputs—can generally be accomplished by the routine application of formulas and mechanical procedures or by the use of a computer. The remaining phases of decision analysis (the selection of model structure and input data, and the use of model outputs) call for judgment and discretion, which cannot be reduced to rote procedures. Management decision making cannot yet, if ever, be accomplished by a set of series of steps which leads to the "one right answer."

Even with the use of formal methods, the decision-making process is not the simple linear process from problem definition to solution that it might appear to be. The formal analysis shows the implications of the input to that model; however, on seeing these implications, the decision maker may wish to revise the input or perhaps even restructure the problem. The solution is likely to be an iterative one involving cycles of model construction and model revision. Formal decision analysis should be seen not as a way of providing some final mathematical solution but rather as a way of modeling the decision problem—articulating the important features, and exploring the implications for action. Its value lies as much in the framework for thinking about complex decisions as it does in the ultimate solution. The focus on "solving" a decision problem using formal procedures often leads to the misimpression that decision analysis is simply a matter of collecting numbers, plugging them into some formulas, and automatically arriving at the optimal act. This in turn leads one to conclude that decision analysis has little connection to real-world decision making. Lest this discussion contribute to this view of decision theory, the reader is urged to keep the broader perspective in mind.

CLASSICAL AND BAYESIAN DECISION MAKING

In this section we examine the classical hypothesis test as a decision procedure and compare it to a Bayesian approach to decision making. In decision making under uncertainty, both the classical and Bayesian approaches must be concerned with acts, unknown states of the world, preferences for consequences, and uncertainty regarding the states. Before looking at hypothesis testing in detail, we briefly review how the two approaches treat these four elements of the decision framework.

THE DECISION FRAMEWORK

Acts and States

In both approaches, the basic structure of the problem is the same. The decision maker wishes to choose among alternative acts, and the consequences of these acts depend upon states that might exist or might occur, but are not known at the time of the decision. In statistical decision problems, the states are the possible values of unknown parameters, and, in this discussion, these parameters express program performance on some outcome measure.

Preferences for Consequences

In the Bayesian approach, the decision maker's preferences for the consequences are expressed in terms of utility, and the existence of a utility function follows from the axioms of coherence. Utility is scaled in a manner similar to that used in scaling subjective probabilities, as was discussed in Chapter 3.

The decision maker's utilities for the possible consequences can be determined as follows: Let c_* be the least preferred consequence and let c^* be the most preferred consequence. Preferences for intermediate consequences are scaled by presenting the decision maker with a choice between receiving an intermediate consequence c for sure, and a lottery that results in c^* with probability P and c_* with probability $(1 - P)$.

Obviously, if P is 1.00, c* is certain to result and the lottery is preferred. On the other hand, if P is 0.00, c_* is certain to result and c for sure is preferred. By adjusting the value of P, we arrive at some point where the decision maker is indifferent between c for sure and the lottery. At that point, the utility for c and the utility for the lottery are equal. The utility of a lottery equals the mathematical expectation of the utility of its "prizes," and we have

$$U(c) = PU(c^*) + (1 - P)U(c_*)$$

Utility is defined only up to a linear transformation and we can arbitrarily set $U(c_*) = 0.00$ and $U(c^*) = 1.00$. We then have

$$U(c) = P(1.00) + (1 - P)(0.00) = P$$

Thus, for any consequence, we can assign a number P between 1.00 and 0.00 that represents the decision maker's utility or relative preference for that consequence.

In the classical approach, it is assumed that the decision maker has some preference function over the possible consequences. The notion of preference in the classical approach is not as precisely defined as it is in the Bayesian approach. The preference function does not follow from any axiomatic basis; it is simply assumed to exist. Consequently, it does not necessarily have to have the same properties as a utility function. On the other hand, a classical decision maker could use the above method for scaling preferences. For comparative purposes, we will take the preference function here to be the same as a utility function. Actually, preference in the classical framework is most often expressed as a loss function. Loss can, however, be interpreted as negative utility, and the same results are obtained whether one works with utility functions or with loss functions.

The subsequent discussion will focus on problems where the unknown state is continuous and the consequences are, therefore, continuous. Here we will express utility as a continuous function of the unknown state. Just as subjective probability distributions can be fit with certain functional forms to simplify the analysis, the same thing can be done with utility functions, and references to this fitting process are provided at the end of the chapter. In this discussion, we will assume that the decision maker's utility function is an increasing function of the uncertain state. In other words, higher reading scores are preferred to lower reading scores, or higher health status is preferred to lower health status, over the range of the uncertain state being considered. More specifically, we will assume that this function is linear over this range. This simplifies the illustrations considerably and does not greatly restrict the conclusions that will be reached. Edwards et al. (1975) base their analysis on linear utility functions and argue that linear functions may be reasonable approximations to nonlinear monotonic functions in many cases.

In many decision problems, the consequences may involve several dimensions, or aspects, of importance to the decision maker. For example, in introducing a new medical treatment, the decision maker's evaluation of the consequences might involve consideration of side effects, costs of new equipment, and patient discomfort, in addition to the cure rate. Assessment of the overall utility of the consequences falls into the area of multiattribute utility theory. There are a number of ways of scaling preferences on the individual dimensions and combining them into an overall utility value. These procedures are described in the reference listed at the end of this chapter and are not discussed here. We will assume in my examples that the utility for the consequences is some linear function of the unknown state plus a fixed amount reflecting some weighted sum of the fixed costs and benefits associated with the act. Because this additional amount is fixed (for a given act), utility will still be expressed as a linear function of the unknown state. In the discussion of the decision problems to follow, I will not describe the actual assessment of utility. The objective here is simply to examine how research findings fit into the decision-making process. Again, the references at the end of this chapter cover the topic of assessment in detail.

Uncertain States

This uncertainty is handled very differently in the two approaches. In the Bayesian approach the decision maker specifies a probability distribution over the states, and then chooses the act that maximizes expected utility with respect to this distribution. If new data are available, the distribution is revised according to Bayes' theorem, and the decision is made using this revised distribution.

In the classical approach, no probabilities are attached to the states. In the situation where no data are available, there are various criteria for choosing among the acts: a "pessimistic" criterion is to select the act that minimizes the maximum loss that could be experienced; an "optimistic" criterion is to choose the act that maximizes the minimum utility that could be experienced. Other criteria are discussed in Luce and Raiffa (1957: ch. 13). The main interest here is in the situation where new data are available, and the decision maker wishes to use the data in making the decision. In the classical approach, one works with decision rules or decision functions that specify which act to select for any specific sample outcome that is observed. The decision rules are evaluated in terms of their associated error probabilities and error costs or loss. Then, rather than choose directly among the acts, the decision maker chooses a decision rule that performs well over the long run with respect to certain criteria concerning error probabilities and loss.

HYPOTHESIS TESTING AND DECISION MAKING

We now turn to classical decision making. Previous chapters have stressed that classical hypothesis testing is a decision procedure: It is a procedure for choosing between hypotheses in light of the cost of error to the decision maker. Furthermore, its use as an inference procedure in evaluation research was questioned. Here the use of hypothesis testing as a decision procedure is examined. It is frequently suggested in classical texts that, if the costs of error were specified, the test could be made directly relevent to a user's decision problem—it could be designed so that acceptance or rejection of the hypotheses would indicate which of two courses of action a user should choose. This notion seems to underlie the suggestion that the user of classical test results make a posteriori adjustments to significance levels in order to bring statistical significance into line with practical significance for acting on the basis of the results.

At this point the use of the classical test in a simple two-action terminal decision problem will be considered. By a "terminal decision" I mean one in which the decision will be made given the information at hand, and that continued experimentation or collection of data is not one of the courses of action under consideration. A hypothesis test would seem to be ideally suited for addressing this type of decision problem. The following discussion will involve consideration of both classical and Bayesian procedures and, in order to avoid switching back and forth between the different notation used for the parameters, the unknown parameters will be symbolized as in the sections on Bayesian inference. The reader should keep in mind that from the classical perspective, these parameters are fixed, rather than random, quantities.

The Decision Problem

Suppose that an evaluation is conducted for a single decision maker who is faced with a changeover decision problem involving two different program models. One model is the program that is presently being used, and an alternative model is a new program that has been proposed. The decision maker wishes to choose between two acts: (a_1) retain the present program, and (a_2) implement the new program, and wishes to make this choice using some measure of program performance that is considered to be of primary importance. An experiment is conducted with random assignment of subjects to two treatment conditions representing the two program models. The performance measure is assumed to be normally distributed within each group with unknown means θ_1 and θ_2 and (for simplification) known variances σ_1^2 and σ_2^2, respectively. The decision maker's preference for the programs depends upon their unknown means θ_1 and θ_2.

This type of decision problem might be addressed using the classical test for the difference between two normal means. The issue is how to conduct the test in such a way that practical and statistical significance will coincide, and that acceptance or rejection of a hypothesis will indicate which action should be chosen. In order to do this, it is necessary to take the cost of error to the decision maker into account. Because we are considering a case in which the test is being designed so that the hypotheses and the acts have a one-to-one relationship, in the following discussion I will refer interchangeably to a choice between the acts and a choice between the hypotheses.

The Cost of Error

The cost of error is usually expressed directly in terms of loss or negative utility. In this discussion, the nature of the cost of error will be clearer if, to start with, we assume that these valuations are expressed in terms of positive utility. As we mentioned above, we will assume that the decision maker's utility is a linear function of the unknown quantity. We also assume utility is assessed for some fixed time period.

The utility functions. Returning now to the decision problem, if the present program is retained (action a_1), the decision maker's utility for the consequences of a_1 as a function of the unknown quantities may be expressed in the linear form

$$U(a_1, \theta_1) = r_1 + s_1\theta_1 \qquad [9.1]$$

The (a_1, θ_1) pair represents a potential consequence when a_1 is chosen. If a_1 is chosen, the utility varies only as a function of θ_1, and θ_2 is not included in the notation. Any fixed component of the utility that does not vary as a function of θ_1 is included in the constant term r_1. Included here would be any fixed costs or benefits incurred in taking the action. The variable component is $s_1\theta_1$, and utility varies directly with the value of θ_1. Similarly, if the new program is implemented (action a_2), the utility function may be expressed as

$$U(a_2, \theta_2) = r_2 + s_2\theta_2 \qquad [9.2]$$

For simplification it will be assumed that the slopes s_1 and s_2 are equal and can be both represented by s. These functions are assumed to be bounded as required by the theory of utility.

The loss functions. Loss is experienced when one chooses an act that has a consequence of less than the maximum utility obtainable when a given state of the world occurs. For example, if $U(a_2, \theta_2) > U(a_1, \theta_1)$ for some particular true values of θ_1 and of θ_2, loss would be experienced if a_1 were chosen. The loss associated with a consequence is obtained by subtracting the utility of that consequence from the maximum utility obtainable with any act under that particular state of the world. In this instance, the loss would be

$$L(a_1, \theta_1, \theta_2) = U(a_2, \theta_2) - U(a_1, \theta_1)$$

Loss is a function of both θ_1 and θ_2. The loss associated with the consequence with the maximum utility is zero because the utility of that consequence subtracted from itself is zero. Loss, as it is used here, is "opportunity loss" and is sometimes called "regret" for obvious reasons.

It can be shown using equations 9.1 and 9.2 that $U(a_2, \theta_2) > U(a_1, \theta_1)$ when

$$\theta_2 - \theta_1 > \frac{r_1 - r_2}{s} \qquad [9.3]$$

We will represent $\theta_2 - \theta_1$ by δ, as we have in the previous chapters. Thus a_2 leads to consequences with the greatest utility, and is the optimal act, when the difference between the means δ exceeds the quantity on the right-hand side of equation 9.3. This right-hand term represents the *break-even value* and will be symbolized by δ_0. It represents the amount by which the mean for the new program must exceed that for the present program, in order for the new program to be considered worth implementing. The difference $r_1 - r_2$ represents differences in the fixed costs and benefits associated with the two actions, and might, for example, reflect differences due to the cost of retraining, reorganization, disruption, new equipment, and so on, in implementing the new program. Dividing by s serves to express this difference in the same units as δ. Similarly, it can be shown that $U(a_1, \theta_1) > U(a_2, \theta_2)$ when δ is less than the break-even value. Thus a_1 leads to consequences with the greatest utility, and is the optimal act when $\delta < \delta_0$.

Consider the loss function when a_1 is chosen. If the true difference between the means is less than the break-even value, then the loss for consequences involving a_1 is zero because a_1 (continuing the present program) is the optimal act. On the other hand, when the true difference is greater than the break-even value, then a_1 is no longer the optimal act and, in this case

$$
\begin{aligned}
L(a_1, \theta_1, \theta_2) &= (r_2 + s_2\theta_2) - (r_1 + s_1\theta_1) \\
&= -(r_1 - r_2) + s(\theta_2 - \theta_1) \\
&= -s\delta_0 + s\delta \\
&= s(\delta - \delta_0)
\end{aligned}
$$

Thus loss depends on how far the true difference is from the break-even value. Similarly, it can be shown that the loss for implementing the new program (action a_2) is zero when $\delta > \delta_0$ and is $s(\delta_0 - \delta)$ when $\delta < \delta_0$.

Rather than work with these separate loss functions for each act, we combine them into one function. The cost of error or loss associated with choosing the wrong action can be written

$$
L(\text{error}, \delta) = s| \delta - \delta_0| \qquad\qquad [9.4]
$$

When $\delta < \delta_0$ the error is implementing the new program (action a_2). When $\delta > \delta_0$ the error is retaining the present program (action a_1). This is a V-shaped function centered on δ_0, and in this problem loss due to the error of selecting the wrong action is zero at the break-even value and increases linearly in either direction.

Solving the Decision Problem

Now that the cost of error is defined, how should the test be conducted to provide an answer to the decision problem? It can be seen from the loss structure that the basic question is whether or not the true difference exceeds the break-even value. If the true difference between the two means exceeds the break-even value, the new program should be implemented; if it does not, the present program should be retained. The most straightforward approach to answering this question would be to test the hypotheses $H_0: \delta \leq \delta_0$ and $H_1: \delta > \delta_0$.

Given these hypotheses, what is the appropriate decision rule? How should the significance level of the test be chosen? Classical texts do not provide much direction here. Most presentations of classical methods for researchers appear to assume that there is one constant cost of error for a Type I error no matter how large it is. The cost of Type II errors is often assumed to be constant also, although in some discussions the cost of Type II error is viewed as being two-valued—H_1 is divided into a region for which errors are "too small to be important" and a region in which they are "important." Consequently, it is not clear how one should proceed in light of the loss structure that we are considering here. When the cost of error depends upon how far the true difference is from the break-even value, it is not obvious how one should balance the error probabilities against the losses. This presents a problem for someone attempting to use the test as a basis for making this terminal decision. The problem, however, is more than just a limitation in the form of the loss function that is considered. Neyman and Pearson (1967: 195-196) do suggest how other loss functions might be taken into account, and subsequent developments of the classical theory, although not widely used by researchers, do allow for more flexibility in the treatment of loss. Yet, as will be seen, difficulties still remain in finding a decision rule that will be optimal for a particular decision problem.

As was mentioned in Chapter 1, Abraham Wald extended the decision theoretic approach of Neyman and Pearson, and provided for a more detailed consideration of loss in his *theory of statistical decision functions*. Examination of the decision problem in these terms highlights a more basic difficulty in the classical to decision making. In this framework, the selection of the appropriate decision rule involves consideration of *risk*. Risk depends on both the specific decision rule that is used and the state of the world. Specifically, it is defined as the average loss of using a given decision rule for a particular state of the world. To understand what risk represents, recall from the earlier discussion that the choice of actions (hypotheses) is determined by a random variable— the test statistic. Given some particular decision rule and the sampling distribution of the test statistic, we can determine the probability of taking either action for given values of δ. Then, from these probabilities and the loss function, risk can be determined. In this problem, the risk in using a particular decision rule at any value of δ reduces to the probability of error at that value of δ, multiplied by the loss due to error of that value of δ.[1]

An illustration of how the risk function is obtained is shown in Figure 9.1. Figure 9.1A shows the conditional probability of error for four different decision rules when testing the hypotheses being discussed. These are error characteristic curves, as discussed in Chapter 2. Figure 9.1B shows the loss due to error as expressed in equation 9.4. Figure 9.1C shows the risk function using the four different decision rules. It can be seen that, for each value of δ, the risk is a product of the error probability and the loss.

The basic idea is to choose a decision rule that minimizes risk. A decision rule that has greater risk over all values of δ than does another rule could be eliminated from consideration. Inspection of the risk functions in Figure 9.1C shows that none uniformly minimizes risk. For example, of these four rules, the one with $\alpha = .01$ has the smallest risk for values of δ less than the break-even value. However, it has extremely large risk for values of δ greater than the break-even value. On the other hand, the rule with $\alpha = .75$ has the smallest risk of the four for values of δ greater than the break-even value, but the largest risk values for δ less than the break-even value. It is interesting that the conventional decision rules are not in any overall sense more stringent than the other rules. Whether we look at the error characteristic curves in Figure 9.1A or at the risk functions in Figure 9.1C, it can be seen that stringency in one region of δ is achieved at the price of lack of stringency in the other regions. The issue of choosing among these decisions rules is not whether or not they are stringent but, rather, where they are stringent.

Intuitively, one's choice of a decision rule should take into account where the true value of δ is likely to lie. That way a decision rule could be chosen that provides the best protection in regions where the true value is likely to occur. However, the frequency interpretation of probability does not allow for formal treatment of these considerations. This is as far as the analysis takes us. We are still left with a set of diverse

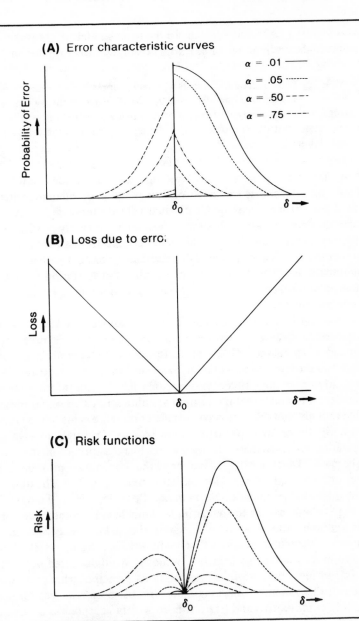

Figure 9.1 Risk Functions for 4 Decision Rules as Determined from the Error Characteristic Curves and the Loss Due to Error

rules—none of which is superior to the others in terms of minimizing risk over the entire range of δ. Raiffa (1968: 277) writes,

> Although Wald's accomplishments were truly impressive, statistical practitioners were left in a quandary because Wald's decision theory did not single out a best strategy but a family of admissible strategies, and in many important statistical problems this family is embarrassingly rich in possibilities.

Although the particular example that we are considering involves linear loss, the choice of a decision rule remains problematic whether loss due to error is linear, quadratic, two-valued, step-wise, or any other form that might reasonably apply. Even with the highly artificial two-valued loss function, the same difficulty arises. Given any of these loss functions, we still obtain a set of decision rules, none of which uniformly minimizes risk. Therefore, even with explicit consideration of loss, the classical approach still does not provide a unique solution to a specific decision problem.

The originators of classical decision methods were well aware of this; however, they saw the choice among decision roles, in practice, as depending upon other considerations that could not be easily incorporated in the theory. Pearson (1962: 55) writes that he and Neyman "were certainly aware that inferences must make use of prior information and that decisions must also take account of utilities, but after some considerable thought and discussion round these points we came to the conclusion, rightly or wrongly, that it was so rarely possible to give sure numerical values to these entities that our line of approach must proceed otherwise." The following section shows that introducing prior probabilities into the analysis provides a solution to this problem. The objective and major achievement of Neyman and Pearson, and of Wald, was to bring the analysis up to this point without having to introduce prior probabilities. Having accomplished this, they acknowledged the importance of other considerations that could include prior probability if the decision maker were so inclined. Actually, Wald discussed both minimax and Bayesian solutions. The minimax principle, which says select the rule that minimizes the maximum risk, has been criticized for being unduly conservative and has not been widely accepted as a basis for selecting good statistical procedures. Although the Bayesian solutions were interesting from a theoretical perspective, neither Wald nor much of his audience was willing to assume that prior probabilities would exist for most statistical problems; Wald's work, however, predated many of the developments in the coherence argument.

Unfortunately, the additional considerations necessary for a solution to the decision problem tend to be ignored, and the classical test is often viewed as providing a full solution. The above discussion shows that this is not the case.

A BAYESIAN SOLUTION

If we introduce prior probability distributions for θ_1 and θ_2, with their implied prior for δ, into this analysis, a solution can be obtained. It is then possible to determine the average risk or *expected risk* for a decision rule, where the risk at any value of δ is weighted by the prior probability of δ. Graphically, the expected risk for a decision rule is the average height of the risk function. The optimal decision rule is that which minimizes the expected risk. Schlaifer (1959: 635) discusses a method for finding the decision rule that minimizes expected risk and it is interesting to note that for this problem, with an approximately uniform prior, it is the rule for which $\alpha = .50$. For a significance level of $\alpha = .50$, the critical value of z for conducting the classical test discussed in Chapter 2 is zero. Using the formula for the test statistic, it is easy to show that H_0 should be rejected, in this case, when $(x_2 - x_1) > \delta_0$. This means that, with noninformative priors, the optimal decision between the two acts could be made simply on the basis of whether or not the difference between the sample means exceeds the break-even value—an elaborate test is hardly necessary here. Inspection of the risk functions in Figure 9.1C shows that the average height (average risk) for a conventional decision rule with $\alpha = .05$ or $\alpha = .01$ is several times that for the rule with $\alpha = .50$, and, as Schlaifer (1959: 633) writes, for problems of this type "tests of significance are completely unnecessary . . . if they are made at level .5 and definitely harmful if they are made at any level which is substantially different from .5."[2]

Of course, different loss functions may lead to different solutions. What is shown here is that, in considering loss functions that are probably a good deal more realistic than the simple two-valued functions usually considered in classical testing, we arrive at a result that differs substantially from recommendations and practice in hypothesis testing. Tests carried out with significance levels of .50 are unheard of. Yet this is exactly what is required if one were to address this particular decision problem using a hypothesis test.

Once one adopts the Bayesian view, a more direct solution to the decision problem is available. Instead of choosing among decision rules, one chooses directly between the actions in such a way as to maximize expected utility. When loss is defined as it is here, selecting the act that maximizes expected utility is equivalent to selecting the act that minimizes expected loss. Thus, rather than attempting to find a decision rule before drawing the sample that will minimize expected risk, one directly chooses the act that minimizes expected loss using the posterior distribution based on the sample that was obtained. Because loss is linear in δ, the Bayesian solution to this problem is fairly simple. It can be shown, by taking the expectations of the loss functions for each act and rearranging, that a_2 will be the optimal act when $E(\delta) > \delta_0$, where $E(\delta)$ is the expected value or mean of the posterior distribution of δ. Thus the choice between a_1 and a_2 can be made simply on the basis of whether or not the posterior mean of δ exceeds the break-even value. If the prior distribution is noninformative, the mean of the posterior distribution of δ is equal to the difference between the two sample means as shown in Table 5.2. Therefore, as above, this decision could be made on the basis of whether or not the difference between the two sample means exceeds the break-even value.[3]

Actually, the Bayesian solution can be even more straightforward than this. Given posterior distributions for θ_1 and θ_2, and the utility functions for each act, all one need do is compute the expected utility of each act and select the act with the maximum expected utility. In making this choice, determination of a break-even value is not necessary. This approach is discussed in more detail in a following section.

The main point here is that the classical hypothesis test by itself has limitations for modeling even very simple decision problems of users of evaluation findings, and providing direction for action. Even if the loss function is specified in detail, there is no satisfactory way within the frequency probability framework of classical statistics of selecting a decision rule that is "best" for a specific decision problem. This is true whether the utility function is linear or any other form that might reasonably apply. It is frequently implied that if the loss structure of the problem were known, the classical hypothesis test could be conducted in such a way that statistical and social significance would coincide and provide direction for action. This is not the case. In order to choose a decision rule and significance level that is appropriate for the decision problem, it is necessary to give some consideration to prior probabilities. If one is willing to do this, then Bayesian inference can be of value, and for purposes of making decisions, much of the framework of hypothesis testing can be dropped in favor of more direct Bayesian procedures.

DECISION MODELING

Bayesian decision theory provides a general framework for modeling decision problems and examining the implications for action. This section shows how a decision problem involving research findings can be described and analyzed. Again, I emphasize that decision analysis is a means for systematically thinking through a decision problem, and is not some automatic decision-making device. As Enthoven (1975: 456) writes, "Good analysis is the servant of judgment, not a substitute for it." In fact, the analysis makes the judgments that are required on the part of the decision maker even more salient than they might otherwise be. The value of decision analysis lies in its "divide and conquer" approach, whereby complex judgments are broken down into simpler judgments, which are then combined in a logical manner to aid in making the complex judgment. Huber (1980: 61-62) points out a number of benefits from the formal analysis: It helps identify inadequacies in the implicit informal models of the problem, it points to new information that may be needed, it serves as an organized external memory of the details of the problem, it enables one to aggregate large amounts of information, and it provides a framework for communication. These benefits may be as important, if not more important, than the ultimate solution that is obtained.

A flexible tool for displaying and analyzing decision problems is a tree diagram. Some basic ideas will be introduced by examining a tree diagram for the simple decision problem that we have been discussing; the diagram is shown in Figure 9.2. Acts and uncertain events or states are shown in forks with separate branches. On the left-hand side, we have a small square with two branches extending out from it. This is an *act fork*—there are branches for the two acts, a_1 and a_2. The names of the acts are indicated on the branches. At the end if each act branch is a small circle with branches extending from it. This is an *event fork*—the branches represent the possible events or, in this case, values of the unknown parameters, that determine the consequences of the act. Because θ_1 and θ_2 are continuous, there is an infinite number of event branches and this is indicated by the dashes between the branches of each fork. When there is a large or an infinite number of branches, this is sometimes called an *event fan*. With a small number of events, the

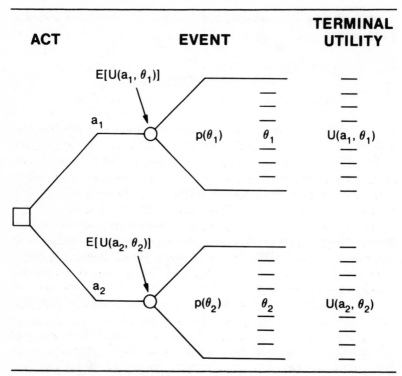

Figure 9.2 A Simple Decision Tree

probability of the event and the name of the event can be listed on each branch. On the right-hand side of this diagram is the utility of arriving at the end points of the decision tree. Each end point in this problem represents a consequence—a combination of a particular act and a particular value of θ_1 or θ_2. Because the utilities are continuous functions of θ_1 and of θ_2, they are simply indicated by the dashes.

The decision problem is solved by starting at the right-hand side and working backward. We compute the expected utility at each event fork and "replace" that fork by a single value—the expected utility. The expected utility is indicated at the end of the arrow pointing to the small circle. We now have a simplified decision problem where the decision maker has to choose between two acts, one of which leads to a single value of $E[U(a_1, \theta_1)]$ and the other to a single value of $E[U(a_2, \theta_2)]$. The decision maker should choose the act with the greatest expected utility.

We note here that, because it is assumed that utility is a linear function of the unknown parameter, in this problem,

$$E[U(a_1, \theta_1)] = E(r_1 + s\theta_1) = r_1 + sE(\theta_1) = U[a_1, E(\theta_1)]$$

and

$$E[(U(a_2, \theta_2)] = E(r_2 + s\theta_2) = r_2 + sE(\theta_2) = U[a_2, E(\theta_2)]$$

where E is the expectation operator. Simply stated, the expected utility is, in this case, the value of the utility function at the expected value, or mean, of the unknown parameter. This means that, for purposes of making this terminal decision, the only features of the posterior distributions of θ_1 and θ_2 that are needed are the means. Instead of having to evaluate an integral to determine expected utility, we work with the utility at the mean of θ_1 or θ_2. This does not mean that the information in the full distribution is being ignored or discarded; rather, the nature of the problem is such that the full expectation reduces to this simple result. Determining expected utilities is therefore greatly simplified when the utility functions are linear.

In some decision problems, one of the alternative acts might involve collecting additional data and examining their implications before making the terminal decision. This is known as a preposterior decision problem because it involves consideration of posterior distributions that might be obtained with the new data. Bayesian decision theory provides methods for determining expected gains in utility or reductions in loss due to possible increased precision from the new data. These considerations are then combined with the cost of additional sampling and with the other features of the decision problem to enable the decision maker to decide whether to collect additional data or to select one of the remaining terminal acts. In contrast to the terminal decision problem, the preposterior decision problem requires the full distribution of the unknown quantities rather than just the means in order to evaluate the expectations involved. In the discussions here, we are considering only terminal decision problems where the decision maker is not considering collecting additional data. The analysis of preposterior decisions is discussed in most decision theory texts such as Raiffa (1968), Schlaifer (1959, 1969), and Winkler (1972). Raiffa and Schlaifer (1961) provide a detailed theoretical treatment of this aspect of decision making. This type of analysis underlies the Bayesian approach to experimental design.

The decision problem diagramed in Figure 9.2 is equivalent to that discussed in the previous section. Because we worked here with utility, rather than loss, this analysis was somewhat more straightforward. In this simple problem, the tree diagram was not necessary; one could easily solve the problem algebraically. The real value of tree diagrams becomes apparent in the analysis of more complex problems. In an example below, a decision involving more than two acts, and multiple uncertainties, is considered.

*   *   *   *   *   *

EXAMPLE 9-1. A DECISION INVOLVING FUNDING FOR AN OUTPATIENT PSYCHIATRIC PROGRAM

A manager of an outpatient psychiatric program has some interest in reorganizing part of the program to provide a new form of day treatment for chronic patients. Research with patients similar to those in the program indicates that the new treatment leads to improved patient functioning when compared to those patients receiving the standard form of treatment. Functioning is assessed on a behaviorally referenced scale involving various activities of daily living; the function scale runs from 0 to 100. On the basis of this research, the manager's beliefs about the mean level of functioning are $\theta_1 \sim N(54.5, 8.02)$ for those receiving the standard form of treatment, and $\theta_2 \sim N(64.1, 9.62)$ for those receiving the new treatment. Those assessed at 60 or above are generally able to maintain part-time employment; those assessed below this level are not, although the score reflects other aspects of functioning as well. (These subjective probability distributions might be data-based distributions that stem from a single study or from multiple studies. Furthermore, they might include both data-based and non-data-based considerations such as were incorporated in the probability assessment procedure described in Example 6-3.) The discussion is simplified here by assuming known variances; the same analysis, however, could easily be carried out when the variances are unknown. The manager's utility for various levels of patient functioning is linearly related to function level, and, in order to make this terminal decision, only the means of the distributions will be needed. The reorganization would require funds for staff training and renovation of some offices. It would also require some additional space, which at present is available in rooms adjacent to the present space. The reorganization would not require additional staff, and it would not change the number of patients that would be seen.

The program manager discovers, on short notice, that funds for one year are available through a state grant program for mental health services. The grant would cover most of the start-up costs, plus certain continuing expenses not included in the present budget. The manager has only a short time to prepare and submit a proposal. There is another review cycle in three months; proposals rejected in the first review cycle will not be considered in the second one. The program manager feels that the proposal that could be prepared for the second review cycle would be considerably more thorough than the one that could be prepared for the first one, and would have a greater chance of being funded. Preparation of this more thorough proposal would, of course, involve more time and effort. The manager would have to start preparing this almost immediately. A drawback to waiting for three months is that the space may not be available. Another service is discussing the possibility of expansion in the next few months, and may move into the vacant space. The manager feels that if the adjacent space were not available, the reorganization would be impractical, and time and effort would have been wasted in preparing a more extensive proposal. The manager does not consider the three-month delay by itself to be a problem; it is the availability of space that is of concern.

The decision involves the acts of whether to apply now, or to begin preparing a more thorough proposal for an application in three months, or simply to do nothing. The manager wants to make this choice, taking into account uncertainties concerning treatment effectiveness, funding, and space, and taking into account the value of improved patient functioning, time and effort devoted to preparing a "quick and dirty" proposal versus a more thorough proposal, and disruption accompanying the reorganization. We will limit the analysis to the problem of obtaining the funding for a one-year period.

A tree diagram of this decision problem is shown in Figure 9.3. Starting on the left, we have an act fork with the three actions just mentioned. If the program manager applies now, the proposal may either be funded or not funded; this is shown in the event fork at the top of the diagram. Depending on the outcome of this funding decision, the manager will either carry out the reorganization or not. It is assumed that if the proposal is funded, the reorganization will be carried out, and if it is not funded the reorganization will not be carried out. The decision forks here have been "pruned" to show that the action is predetermined

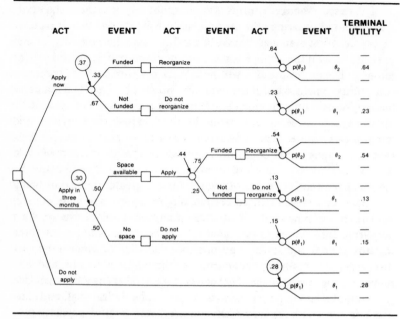

Figure 9.3   Decision Tree for Application for Funding Problem

given the outcome of the event fork. Moving over to the right-hand side of the diagram, the terminal event forks that express uncertainty about the mean level of patient functioning with and without the new treatment are shown. The manager's terminal utilities are linearly related to the average level of functioning of patients to be treated during the year, with and without the new treatment, and also reflect any fixed costs of proposal preparation and disruption due to reorganization associated with the acts leading to a particular terminal branch. For purposes of calculating expected utility with these linear functions, we only need to know the value of the utility function at the expected value or mean of the unknown parameter, and we show only the utilities at the values $\theta_1 = 54.5$ and for $\theta_2 = 64.1$ under the Terminal Utility heading.

Going back to the left-hand side of the figure again, we see that if the manager waits, the next step will depend on whether or not space is available. This is shown as an event fork on the middle branch. If space is available, then the manager will submit the proposal and apply for funding. This is followed by another event fork expressing uncertainty about funding. The interpretation of the remaining branches is similar.

The decision maker's subjective probabilities concerning the uncertain events of funding and the availability of space are shown on the respective event branches. We see that the manager believes the probability of funding if the application is submitted now is .33, whereas with a more thorough proposal it will be .75. Also, the manager's subjective probability that space will be available is .50. The uncertainties about patient functioning with the two types of treatment are expressed in the distributions for θ_1 and θ_2, and the relevant information has been incorporated in the expected values as discussed above.

The problem is solved by continuing to work backward from the right side of the diagram, replacing the event forks by their expected utilities, and choosing the branches of the act forks with the greatest expected utilities. Beginning on the right-hand side, the event forks involving θ_1 and θ_2 are replaced by the utilities of the expected values. These in turn are used in calculating the expected utilities at preceding event forks. Finally, the decision maker's problem is reduced to choosing among acts that have the expected values enclosed in circles. The alternative of applying now has the highest expected utility (.37) and is the optimal act.

At this point the decision problem is, in a sense, "solved." However, the decision maker might well ask some "what if" questions to determine how sensitive the solution is to the specification of subjective probabilities and utilities. For example, what would happen if the program manager's probability of receiving funding with the immediate application were only .25, or .20? Would this change the decision? The answer is no. This probability would have to be .17 or less before the act of applying now would be suboptimal. Thus, as long as the manager feels that this probability is well above .17, the choice of actions will be robust with respect to minor variations in this probability.

As another example, suppose the mean of θ_2 were only 59.3; the difference between the means of θ_1 and θ_2 is then only half of what it was considered to be originally. The terminal utilities at the first and third terminal branches are then .38 and .28, respectively. In this case the expected utility of applying now is .28, the expected utility of applying in three months is .20, and the expected utility of not applying remains .28. Under these conditions the act of applying now is no better than the act of not applying at all. However, as long as the mean of θ_2 is considered by the decision maker to be larger than 59.3, the act of applying now is the optimal act. Other quantities could be varied also, singly or in combination, to see how changes might affect the decision. (Computer programs such as that described in Lau et al., 1983, provide an efficient means of doing this and displaying the implications.) Examination of the results may also suggest ways of reconceptualizing the decision problems—there may be other actions and events that need to be included in the model. A thorough analysis might go through several iterations.

A comment on group decision making is necessary. Suppose that the program manager were working with members of the clinical team that would be affected by the reorganization, and that they were trying to reach some group decision on which action to take. If the group members agreed on the probabilities and utilities, there would be no problem—they could analyze their decision problem using these consensus values. If there were disagreements, the problem could be analyzed separately for the disagreeing individuals. It is possible that there can be disagreements on some points and yet the different parties may still arrive at the same decision. In some cases, the disagreements will lead to different decisions; however, the analysis can still be helpful. Suppose that there were disagreement about the probability of funding. The analysis could help pinpoint the source of the disagreement here, and further discussion might resolve it. Still, after all this, there may be remaining disagreements. At this point, neither the Bayesian approach nor any other approach can magically bring about agreement where it does not exist. Such disagreements would have to be resolved through traditional processes for dealing with conflicting interests, such as negotiation or exercise of authority. Of course, in adversarial settings there may be obvious disagreements at the outset and the different parties may have no interest in revealing their beliefs and preferences to their adversaries. Yet decision analysis can still be of value to the individuals involved in personally deciding among the actions available to them, while taking the actions of their adversaries into account.

* * * * * *

Bayesian decision theory provides a framework for thinking through complex decision problems involving various types of uncertainty. In this problem, one source of uncertainty is that pertaining to program effectiveness; and the inferences from the research, expressed in posterior distributions, are in the exact form that is needed to take this uncertainty into account in decision analysis. It is hard to see how a significant/nonsignificant answer provided by a classical hypothesis test could be of direct value in helping the program manager analyze this decision problem. Because the classical analysis is restricted to probabilities that have long-run frequency interpretations, the uncertainties concerning space and funding could not be given formal consideration in any classical treatment of the decision problem. The Bayesian approach provides a way of displaying a great deal of information and examining the implications. It enables one to reduce a complex problem to a number of simpler components and then to combine them in a coherent fashion in selecting a course of action. There is, of course, some effort involved in carrying out this type of analysis, and for very simple problems, it would hardly be worth the effort. The point is that, when a decision maker is faced with a complex decision problem with significant consequences, a technology is available to aid decision making.

FURTHER READING

The theory of statistical decision functions is presented in Wald (1950). Atchison (1970) and Chernoff and Moses (1959) provide an introductory treatment of this topic, and this topic is considered in statistical theory texts such as Lindgren (1976). Ferguson (1976) reviews the development of this approach. There are a number of useful introductions to Bayesian decision modeling; these include Behn and Vaupel (1982), Brown et al. (1974), Huber (1980), Raiffa (1968), and Schlaifer (1969). Modeling of multiattribute decision problems is discussed in Edwards et al. (1975), Edwards and Newman (1982), Huber (1980), Keeney and Raiffa (1976), and Pitz and McKillip (1984). Assessment procedures for subjective probabilities and utilities are discussed in a number of the above references. Thompson (1982) applies decision theory to decisions about conducting evaluations. Chen and Novick (1982), Novick and Lindley (1978, 1979), and Novick et al. (1980a) discuss fitting utilities with certain functional forms; some of those assessment procedures are available in the CADA package. Hull et al. (1973) classify methods for assessing utility, and Farquhar (1984) provides a recent review of this area. Brown and Lindley (1982) discuss the resolution of incoherence. Recent comments on decision theory and brief overviews can be found in DeGroot (1982), Edwards (1984), Fischhoff (1980), Freeling (1984), Howard (1980), Keeney (1982), McNeil and Pauker (1984), and Winkler (1982b). The decision theory references cited in Chapter 3 are relevant here also. Bayesian statistical decision theory is covered in DeGroot (1970), LaValle (1970, 1978), Pratt et al. (1965), Raiffa and Schlaifer (1961), Schlaifer (1959, 1961), and Winkler (1972).

NOTES

1. Risk as a function of decision rule D and state of the world δ is

$$R(D, \delta) = P(\text{choose } a_1 | D, \delta)L(a_1, \delta) + P(\text{choose } a_2 | D, \delta)L(a_2, \delta)$$

Because loss is zero when the action is the optimal one for a given value of δ, this can be simplified to

$$R(D, \delta) = P(\text{error} | D, \delta)L(\text{error}, \delta)$$

2. Sometimes it is argued that when the break-even value is greater than zero, testing at zero is appropriate using conventional decision rules. The argument is made that the conventional decision rule in this case provides protection against concluding that there is a difference between the treatment means when the true difference is small and practically insignificant. Although it is true that the conventional decision rule will provide some protection in this regard, the use of arbitrary significance levels will almost certainly result in too much or too little protection. Pratt (1965: 187) comments that this test will result in a good procedure only through some remarkable coincidence or remarkable intuition in the choice of a significance level. Even if this did provide the desired amount of protection, asking whether or not δ exceeds zero is a rather roundabout way of addressing the question of whether or not δ exceeds the break-even value.

3. To simplify the argument here, it was assumed that the variances were known. A similar argument can be made when the variances are unknown, but equal, for example. Following Raiffa and Schlaifer (1961: 35), we note that the unknown common variance is a nuisance parameter in that it is irrelevant to the utility or loss of the consequences and enters the decision only through its effects on the probability of the observations. After the observations are obtained, the choice of a terminal act depends not upon the complete joint distribution of the means and common variance but only on the marginal posterior distribution of δ. Given this marginal posterior distribution, we can compute the expected loss of each act and select the one that minimizes expected loss.

10

Summary and Perspective

In concluding this book, we summarize some of the main advantages of the Bayesian approach and suggest some developments that would facilitate the application of Bayesian methods. We then mention two issues that are useful in putting the classical and Bayesian approaches into perspective.

THE MAJOR ADVANTAGES

The advantages of the Bayesian approach for evaluation research fall into three general areas. First, the results of the Bayesian analysis are more direct and intuitively useful than the results of the classical analysis. The probabilities expressed in a Bayesian posterior distribution refer directly to the unknown quantity of interest, whereas the probabilities associated with the results of the classical methods described here refer to the sample statistics and indicate how the method would perform over the long run. A posterior distribution directly expresses uncertainty about the parameter of interest; it graphically shows the final precision resulting from the evaluation. The significant/nonsignificant result of a hypotheses test obscures this uncertainty. In this respect, subjective probability provides a precise and useful language for communicating prior and posterior uncertainties about outcome measures.

Even if classical and Bayesian analyses yield results that are numerically similar, the meaning and correct interpretation of the results are, as was emphasized, quite different. Classical results are, however, often misinterpreted. Raiffa and Schlaifer (1961: 69) write, "Unsophisticated users of statistical reports usually *do* interpret probability statements about statistics as if they were probability statements about parameters because their intuition tells them that probability statements about parameters are directly relevant to their problem while probability statements about statistics are not." In the Bayesian view, classical methods have worked because when misinterpreted, the results often correspond to Bayesian results.

Second, Bayesian methods provide a way of taking prior information into account in inference and decision making. Jackson (1976: 7-8) writes, "if I say to an experimenter: 'What you know after completing an experiment comprises what you knew beforehand and what you learnt from the experiment' he will wonder why I waste time uttering such a platitude"; yet, he goes on to point out, Bayes' theorem is simply a mathematical formulation of that statement. Prior information does and should affect the interpretation of the findings. Different individuals with different prior beliefs may interpret the data differently—the data do not necessarily "speak for themselves." If the purpose of an evaluation is to bring about some agreement regarding the performance of a program, then the study must be designed to have sufficient precision to accomplish this. If the study is weak in relation to prior information, its impact will be diminished. If may not be possible to express prior information in terms of long-run frequency-based probability for incorporation in classical analyses; however, this does not mean that it can be ignored. In contrast to classical methods, Bayesian methods enable one to use all relevant information whether it comes from the sample or from other sources. When the amount of prior information is small in relation to that contained in the sample, it may be sufficient to express this prior information in the form of a noninformative prior distribution. If, however, the amount of prior information is substantial in relation to the sample information, consideration of an informative prior is appropriate and Bayes' theorem provides a mechanism for combining these two types of information. Even when prior information is vague, there are still substantial advantages to be gained by expressing this as an imprecise prior and then applying Bayes' theorem to yield a posterior distribution for the quantity of interest.

Third, the results of Bayesian inference are in the exact form needed for decision making when this is desired. Bayesian decision theory specifies how this information should be used in making a decision. It provides a flexible framework for looking at the implications of the research findings along with various other sources of uncertainty that might be relevant to the user's decision problem. The important role of the user's values in this is clearly spelled out. Inference and decision making are clearly distinguished within the Bayesian framework, and the issue of practical versus statistical significance does not arise. If an individual simply wants to use the results to guide his or her thinking, then no decision is necessary—a posterior is the relevant expression of modified belief. If the user does have a true decision problem, Bayesian decision theory provides a way of addressing that decision.

All three of these areas have direct implications for the utilization of evaluation research. Bayesian statistics directly answers the user's questions about what to think or do on the basis of new data. Furthermore, the approach has a normative basis—it spells out how the results should be used. If appropriate utilization of evaluation findings is of concern, then the Bayesian approach deserves consideration.

DEVELOPMENTS TO AID APPLICATION

At this point, I mention some developments that would facilitate application of Bayesian methods. Specifically, I mention developments that applied researchers could contribute to, or support, in the course of their research. First, there is need for further work on methods of assessing subjective probabilities and utilities, and for evaluation of how well these methods work in actual applications. Social scientists have long been interested in measuring subjective variables, and in the quality of such measurement. These interests are directly relevant to assessment problems. Developments in this area, especially those linked to interactive computerized procedures, may contribute as much as anything to the acceptance and use of Bayesian methods. Second, development of procedures and guidelines for reporting Bayesian results would be of value. There is considerable opportunity here for developing innovative graphical and tabular displays to make the implications of posterior distributions clear, and for testing displays with different kinds of audiences. Especially important here would be the development of methods for concisely displaying the implications of different priors. This would enable the user to assess the robustness of the Bayesian results over a range of prior distributions. Third, some well-documented studies using Bayesian and classical methods as competing methods of analysis would be helpful. Ideally, the documentation accompanying such studies would include discussion of differences or similarities in the results, time and effort required to carry out the analyses, and user's reactions to the different types of results provided. Interested researchers could then judge the relative merits of the two approaches in actual applications.

BAYESIAN DIVERSITY

Throughout the discussion we have referred to *the* Bayesian approach. Actually, there is no one Bayesian approach. As in any scientific field, there are differences in interpretation of various aspects of theory, and differences in notions of how to proceed on various problems. (A glance at the discussion accompanying the papers in Bernardo et al., 1980, will bear this out.) One difference, for example, was mentioned in Chapter 3; there the distinction was made between Bayesians who accept the logical interpretation of probability and those who accept the subjective interpretation. The subjective interpretation was adopted in this book because it leads directly into a theory of decision making. There are other differences as well. Good (1971) in a partly tongue-in-cheek letter, classifies Bayesians on eleven facets or dimensions, and arrives at "46656 varieties of Bayesians." (As this far exceeds the number of Bayesians, there are obviously a number of empty cells.) The problem of distinguishing among the different types of Bayesians has led one wag to comment, according to Geisser (1969), "Ye shall know them by their posteriors." Actually, many of these differences have more to do with choice of axioms, domain of application, and so on than they do with differences in obtained posterior distributions. Yet, when compared to classical statistics, with its diverse collection of principles and procedures, Bayesian statistics constitutes a relatively unified theory. For all Bayesians, inference is based on the single principle of revising priors according to Bayes' theorem, and for those Bayesians addressing problems of decision making under uncertainty, choice is based on the single principle of maximizing expected utility.

CLASSICAL AND BAYESIAN
METHODS IN PERSPECTIVE

This book has dealt with classical and Bayesian methods as alternative methods of analysis for inference and decision problems. Indeed, much of the debate over the merits of classical and Bayesian methods has focused on how the same problems are addressed in each approach. However, it can be useful to stand back and look at this contrast in a broader framework. Leamer (1978) points out that data analysis can be regarded to include three distinct phases. In the first, the data evidence is summarized. In the second phase, it is interpreted, and in the third phase, decisions may be made. The first phase requires a theory of sufficient statistics so that the import of the full set of data can be conveyed through a small number of statistics. The second requires a theory of learning, and the third requires a theory of decision making under uncertainty. The Bayesian approach formally addresses all three phases. The classical approach, in contrast, lacks a formal interpretation phase and, as we have argued, the classical theory of decision is incomplete. Leamer points out that classical inference is basically only a method of data summarization.

In these terms, it can be argued that many of the problems that have been discussed arise when classical procedures are presented, or interpreted, as learning or decision procedures. This leads to an overemphasis on the data, to the exclusion of other information. While it is reasonable to focus on the data in the summary phase, the learning and decision phases necessarily require consideration of prior information. Furthermore, this also leads to a false sense of objectivity. While the data summary phase may be relatively objective, the learning and decision phases must involve subjective considerations. Ignoring these subjective elements in the latter phases does not make the analysis objective; if anything, it makes the analysis irrelevant to the user's questions. If the user of the data has a decision problem, for example, it is hard to conceive of what an "objective" decision might be, other than one in which subjective considerations are explicitly introduced so that a choice can be made in a coherent manner.

Actually, the objective/subjective distinction might be better viewed as a continuum of intersubjectivity. To begin the analysis, the data analyst must choose some probability model to describe the population or process of interest. This model is a very specific statement about the data-generating process and is often based to a large extent on the analyst's judgments and subjective considerations. Indeed, Bayesians have argued that choosing a probability model for the population or process is no more objective than choosing a probability model to represent prior information. What makes the former choice seem to be more objective is that we are often likely to agree on the subjective considerations that underlie this choice. Indeed, for the data summary phase to be of much value to a wide audience, there must be agreement at this stage because sufficient statistics are sufficient only in relation to a specific probability model. Perhaps the only truly objective presentation of the results would be to simply list the raw data. In the learning and decision phases intersubjectivity may not exist to the degree that it does in the data summary phase because different individuals may have different prior information and different preferences. If an investigator is conducting the study to guide his or her thinking or decision making, this issue may not be important. However, if the investigator hopes to convince others, then either he or she needs to convince them that his or her subjective considerations are appropriate, or to design the study in such a way that the sample will be sufficiently precise to minimize the effects of subjective differences.

In this light, many of the differences between the classical and Bayesian approach arise, not because the two approaches have different ways of doing the same thing but because the formal scope of Bayesian statistics is much greater than that of classical statistics. In the Bayesian approach, the data are summarized in the likelihood and the sufficient statistics. In addition, Bayesian statistics formally prescribes how learning from the data and decision making should be carried out; the classical approach does not. For this reason, it is more directly useful in answering the user's questions. However, this comes with a cost—priors and utilities must be specified. For relatively unimportant problems we may not want to devote much effort to such specifications. On the other hand, if the problem is important and the cost of error is great, we may need to devote the effort here. This may require some hard thinking; there is no denying this. Yet as Berger (1980: 58) writes, "It may be hard, but it's the only game in town besides inspired guesswork."

Of course, the classical approach does not deny the relevance of these subjective considerations. However, they are seen as external to the formal analysis; they are to be introduced informally by the researcher or user of the results. The difficulty here is that there is no framework for making these informal considerations explicit and examining their implications, and, in practice, they are often ignored altogether. Therefore, as was noted at the beginning of this book, the Bayesian approach can be seen in a general way as a formal completion of the classical approach, and its advantages for evaluators lie in the fact that it makes explicit the subjective considerations that must be involved in the analysis of data and use of research findings.

References

AITCHISON, J. (1970) Choice Against Chance: An Introduction to Statistical Decision Theory. Reading, MA: Addison-Wesley.

ALLAIS, M. (1953) "Le comportement de l'homme rationnel devant le risque: critique des postulats et axiomes de l'ecole Americaine." Econometrica 21: 503-546.

———and O. HAGEN [eds.] (1979) Expected Utility Hypotheses and the Allais Paradox: Contemporary Discussions of Decisions under Uncertainty with Allais' Rejoinder. Dordrecht, Holland: D. Reidel.

American Statistical Association and the Institute of Mathematical Statistics [eds.] (1981) The Writings of Leonard Jimmie Savage: A Memorial Selection. Washington, DC: Author.

ANSCOMBE, F. J. (1963) "Sequential medical trials." Journal of American Statistical Association 58: 365-383.

———(1961) "Bayesian statistics." American Statistician 15: 21-24.

BAKAN, D. (1970) "The test of significance in psychological research," pp. 231-251 in D. E. Morrison and R. E. Henkel (eds.) The Significance Test Controversy. Chicago: Aldine. (Reprinted from Bakan, D. (1967) On Method, ch. 1. San Francisco: Jossey-Bass.)

BARNARD, G. A. (1958) "Thomas Bayes—A biographical note." (and reproduction of Bayes, 1763) Biometrika 45: 293-315.

BARNETT, V. (1982) Comparative Statistical Inference. London: John Wiley.

BAYES, T. (1958) "An essay towards solving a problem in the doctrine of chances." Philosophical Transactions of the Royal Society 53: 370-418. (orig. pub. 1763; Reproduced with biography in G. A. Barnard, 1958, "Studies in the history of probability and statistics: IX." Biometrika 45: 293-315)

BEACH, L. R., J. CHRISTENSEN-SZALANSKI, and V. BARNES (1984) "On the quality of human judgment: bad news or bad press?" Presented at the Fourth International Symposium on Forecasting, London, England, July.

BEACH, L. R., P. HUMPHREYS, O. SVENSON, and W. A. WAGENAAR [eds.] (1980) Acta Psychologica. (Special Issue) 45.

BEHN, R. D. and J. W. VAUPEL (1982) Quick Analysis for Busy Decision Makers. New York: Basic.

BERGER, J. O. (1984) "The robust Bayesian viewpoint," pp. 63-124 in J. B. Kadane (ed.) Robustness of Bayesian Analyses. Amsterdam, Holland: Elsevier Science.

———(1980) Statistical Decision Theory: Foundations, Concepts, and Methods. New York: Springer-Verlag.

BERKELEY, D. and P. HUMPHREYS (1982) "Structuring decision problems and the 'bias heuristic'." Acta Psychologica 50: 201-252.

BERNARDO, J. M. (1979) "Reference posterior distributions for Bayesian inference." Journal of Royal Statistical Society (B) 41: 113-147.

———M. H. DeGROOT, D. V. LINDLEY, and A.F.M. SMITH (1980) Bayesian Statistics: Proceedings of the First International Meeting held in Valencia, Spain, May 28 to June 2, 1979. Valencia, Spain: University Press.

BIRNBAUM, A. (1978) "Likelihood," pp. 519-522 in W. H. Kruskal and J. M. Tanur (eds.) International Encyclopedia of Statistics. New York: Free Press.

———(1962a) "Another view on the foundations of statistics." American Statistician 16: 17-21.

————(1962b) "On the foundations of statistical inference." Journal of the American Statistical Association 57: 269-326.

BOX, G.E.P. (1980) "Sampling and Bayes' inference in scientific modelling and robustness." Journal of the Royal Statistical Society (A) 143: 383-430.

————and G. C. TIAO (1973) Bayesian Inference in Statistical Analysis. Reading, MA: Addison-Wesley.

BROEMLING, L. D. (1985) Bayesian Analysis of Linear Models. New York: Marcel Dekker.

BROWN, R. V., A. S. KAHR, and C. PETERSON (1974) Decision Analysis for the Manager. New York: Holt, Rinehart & Winston.

BROWN, R. V. and D. V. LINDLEY (1982) "Improving judgment by reconciling incoherence." Theory and Decision 14: 113-132.

CHEN, J. J. and M. R. NOVICK (1982) "On the use of a cumulative distribution as a utility function in educational or employment selection." Journal of Educational Statistics 7: 19-35.

CHERNOFF, H. and L. E. MOSES (1959) Elementary Decision Theory. New York: John Wiley.

CHRISTENSEN-SZALANSKI, J.J.J. and L. R. BEACH (1984) "The citation bias: fad and fashion in the judgment and decision literature." American Psychologist 39: 75-78.

COHEN, L. J. (1981) "Can human irrationality be experimentally demonstrated?" Behavioral and Brain Sciences 4: 317-370 (Additional discussion in 6: 487-533).

CORNFIELD, J. (1970) "The frequency theory of probability, Bayes' theorem, and sequential clinical trials," pp. 1-28 in D. L. Meyer and R. O. Collier, Jr. (eds.) Bayesian Statistics. Itasca, IL: F. E. Peacock.

————(1966) "Sequential trials, sequential analysis and the likelihood principle." American Statistician 20: 18-23.

————(1969) "The Bayesian outlook and its application." Biometrics 25: 617-657 (with discussion).

COX, D. R. and D. V. HINKLEY (1974) Theoretical Statistics. London: Chapman and Hall.

CRANE, J. A. (1982) The Evaluation of Social Policies. Boston: Kluwer-Nijhoff.

DEELY, J. J. and D. V. LINDLEY (1984) "Bayes empirical Bayes." Journal of American Statistical Association 76: 833-841.

DE FINETTI, B. (1978) "Probability: interpretations," pp. 744-754 in W. H. Kruskal and J. M. Tanur (eds.) International Encyclopedia of Statistics. New York: The Free Press.

————(1974a) Theory of Probability: A Critical Introductory Treatment, Vol. 1. London: John Wiley. (translated from Teoria Delle Probabilita (1970))

————(1974b) "Bayesianism: its unifying role for both the foundations and applications of statistics." International Statistical Review 42: 117-130.

————(1972) "Probability, statistics and induction: their relationship according to the various points of view," pp. 147-227 in B. De Finetti (ed.) Probability, Induction and Statistics: The Art of Guessing. London: John Wiley. (translated from "La probabilita e la statistica nei rapporti con l'induzione, secondo i diversi punti di vista." Centro Internazionale Matematico Estivo. Cremonese Rome, 1959)

————(1964) "Foresight: its logical laws, its subjective sources," pp. 93-158 in H. E. Kyburg, Jr., and H. E. Smokler (eds.) Studies in Subjective Probability. New York: John Wiley (translated from Annales de l'Institut Henri Poincare, vol. 7, 1937)

DEGROOT, M. H. (1982) "Decision theory," pp. 277-286 in S. Kotz and N. L. Johnson (eds.) Encyclopedia of Statistical Sciences (vol. 2). New York: John Wiley.

————(1970) Optimal Statistical Decisions. New York: McGraw-Hill.

DICKEY, J. (1973) "Scientific reporting and personal probabilities: Student's hypothesis." Journal of the Royal Statistical Society (B) 35: 285-305.

EDWARDS, A.W.F. (1972) Likelihood: An Account of the Statistical Concept of *Likelihood* and Its Application to Scientific Inference. London: Cambridge University Press.

EDWARDS, W. (1984) "Decision analysis: a nonpsychological psychotechnology," pp. 341-353 in V. Sarris and A. Parducci (eds.) Perspectives in Psychological Experimentation: Toward the Year 2000. Hillsdale, NJ: Lawrence Erlbaum.

————(1983) "Human cognitive capabilities, representativeness, and ground rules for research," pp. 507-513 in P. Humphreys et al. (eds.) Analysing and Aiding Decision Processes. Amsterdam, Holland: North-Holland.

————and M. GUTTENTAG (1975) "Experiments and evaluations: a reexamination," pp. 409-463 in C. A. Bennett and A. A. Lumsdaine (eds.) Evaluation and Experiment. New York: Academic.

EDWARDS, W. and J. R. NEWMAN (1982) Multiattribute Evaluation. Beverly Hills, CA: Sage.

EDWARDS, W., M. GUTTENTAG, and K. SNAPPER (1975) "A decision-theoretic approach to evaluation research," pp. 139-181 in E. L. Struening and M. Guttentag (eds.) Handbook of Evaluation Research (vol. 1). Beverly Hills, CA: Sage.

EDWARDS, W., H. LINDMAN, and L. D. PHILLIPS (1965) "Emerging technologies for making decisions," pp. 259-325 in T. M. Newcomb (ed.) New Directions in Psychology II. New York: Holt, Rinehart & Winston.

EDWARDS, W., H. LINDMAN, and L. J. SAVAGE (1963) "Bayesian statistical inference for psychological research." Psychological Review 70: 183-242.

EELLS, E. (1982) Rational Decision and Causality. London: Cambridge University Press.

EINHORN, H. J. and R. M. HOGARTH (1981) "Behavioral decision theory: processes of judgment and choice," pp. 53-88 in M. R. Rosenzweig and L. W. Porter (eds.) Annual Review of Psychology, vol. 32. Palo Alto, CA: Annual Reviews.

ELLSBERG, D. (1961) "Risk, ambiguity, and the Savage axioms." Quarterly Journal of Economics 75: 643-669.

ENTHOVEN, A. C. (1975) "Ten practical principles for policy and program analysis," pp. 456-465 in R. Zeckhauser et al. (eds.) Benefit-Cost & Policy Analysis 1974. Chicago: Aldine.

FAIRWEATHER, G. W. and L. G. TORNATZKY (1977) Experimental Methods for Social Policy Research. New York: Pergamon.

FARQUHAR, P. H. (1984) "Utility assessment methods." Management Science 30: 1283-1300.

FEATHER, N. T. [ed.] (1982) Expectations and Actions: Expectancy-Value Models in Psychology. Hillsdale, NJ: Lawrence Erlbaum.

FELLNER, W. (1961) "Distortion of subjective probabilities as a reaction to uncertainty." Quarterly Journal of Economics 75: 670-689.

HILDRETH, C. (1963) "Bayesian statistics and remote clients." Econometrica 31: 422-438.

HOGARTH, R. M. (1980) Judgement and Choice: The Psychology of Decision. Chichester: John Wiley.

HOLLAND, J. D. (1962) "The Reverend Thomas Bayes, F.R.S. (1702-61)." Journal of the Royal Statistical Society (B) 125: 451-461.

HOULE, A. (1978) "Exploring the world of Bayesian statistics" (written in French). Proceedings of the Social Statistics Section, American Statistical Association.

FENNESSEY, J. (1977) "Improving inference for social research and social policy: The Bayesian paradigm." Social Science Research 6: 309-327.

———(1976) "Social policy research and Bayesian inference," pp. 269-282 in C.C. Abt (ed.) The Evaluation of Social Programs. Beverly Hills, CA: Sage.

———(1972) "Some problems and possibilities in policy-related social research." Social Science Review 1: 359-383.

FERGUSON, T. S. (1976) "Development of the decision model," pp. 335-346 in D. B. Owen (ed.) On the History of Statistics and Probability. New York: Marcel Dekker.

FINE, T. L. (1973) Theories of Probability: An Examination of Foundations. New York: Academic.

FISCHHOFF, B. (1980) "Clinical decision analysis." Operations Research 28: 28-43.

FISHBURN, P. C. (1981) "Subjective expected utility: a review of normative theories." Theory and Decision 13: 139-199.

———(1970) Utility Theory for Decision Making. New York: John Wiley.

FISHER, R. A. (1973) Statistical Methods and Scientific Inference. New York: Hafner.

———(1970) Statistical Methods for Research Workers. New York: Hafner.

———and F. YATES (1963) Statistical Tables for Biological Agricultural and Medical Research. Edinburgh: Oliver and Boyd.

Foundations of Probability and Statistics (1977) Special issue, Synthese 36.

FREELING, A.N.S. (1984) "A philosophical basis for decision aiding." Theory and Decision 16: 179-206.

GEISSER, S. (1969) "Discussion of J. Cornfield's paper." Biometrics 25: 643-645.

GOKHALE, D. V. and S. J. PRESS (1982) "Assessment of a prior distribution for the correlation coefficient in a bivariate normal distribution." Journal of the Royal Statistical Society (A) 145: 237-249.

GOOD, I. J. (1983) "Some logic and history of hypothesis testing," pp. 129-148 in I. J. Good (ed.) Good Thinking: The Foundations of Probability and Its Applications. Minneapolis: University of Minnesota Press. (Reprinted from J. C. Pitt [ed.] (1981) Philosophical Foundations of Economics. Dordrecht, Holland: D. Reidel)

———(1982) "Axioms of probability," pp. 169-176 in S. Kotz and N. L. Johnson (eds.) Encyclopedia of Statistical Sciences (vol. 1). New York: John Wiley.

———(1978) "A. Alleged objectivity: a threat to the human spirit?" International Statistical Review 46: 65-66.

———(1971) "46656 varieties of Bayesians." American Statistician 25: 62-63.

———(1965) The Estimation of Probabilities: An Essay on Modern Bayesian Methods. Cambridge, MA: MIT Press.

———(1959) "Kinds of probability." Science 129: 443-447.

———(1950) Probability and the Weighing of Evidence. London: Charles Griffin.

GREENWALD, A. G. (1975) "Consequences of prejudice against the null hypothesis." Psychological Bulletin 82: 1-20.

HAMAKER, H. C. (1977) "Bayesianism: a threat to the statistical profession?" International Statistical Review 45: 111-115.

HAMPTON, J. M., P. G. MOORE, and H. THOMAS (1973) "Subjective probability and its measurement." Journal of the Royal Statistical Society (A) 136: 21-42.

HARTIGAN, J. A. (1983) Bayes' Theory. New York: Springer-Verlag.

HAYS, W. L. (1973) Statistics for the Social Sciences. New York: Holt, Rinehart & Winston.

———and R. L. WINKLER (1971) Statistics: Probability, Inference, and Decision. New York: Holt, Rinehart & Winston.

HOWARD, R. A. (1980) "An assessment of decision analysis." Operations Research 1: 4-27.

HUBER, G. P. (1980) Managerial Decision Making. Glenview, IL: Scott, Foresman.

HULL, J., P. G. MOORE, and H. THOMAS (1973) "Utility and its measurement." Journal of the Royal Statistical Society (A) 136: 226-247.

HUMPHREYS, P., O. SVENSON, and A. VARI [eds.] (1983) Analysing and Aiding Decision Processes. Amsterdam, Holland: North-Holland.

HUNTER, D. E. (1984) Political/Military Applications of Bayesian Analysis: Methodological Issues. Boulder, CO: Westview.

IVERSON, G. R. (1970) "Statistics according to Bayes," pp. 185-199 in E. F. Borgatta and G. W. Bohrnstedt (eds.) Sociological Methodology 1970. San Francisco: Jossey-Bass.

JACKSON, P. H. (1976) "The philosophy and methodology of Bayesian inference," pp. 3-16 in D.N.M. de Gruijter and L.J.Th. van der Kamp (eds.) Advances in Psychological and Educational Measurement. London: John Wiley.

———M. R. NOVICK, and D. F. DEKEYREL (1980) "Adversary preposterior analysis for simple parametric models," pp. 113-132 in A. Zellner (ed.) Bayesian Analysis in Econometrics: Essays in Honor of Harold Jeffreys. Amsterdam: North-Holland.

JAYNES, E. T. (1976) "Confidence intervals vs. Bayesian intervals," pp. 175-257 (with discussion) in W. L. Harper and C. A. Hooker (eds.) Foundations of Probability Theory, Statistical Inference and Statistical Theories of Science: Vol 2, Foundations and Philosophy of Statistical Inference. Dordrecht, Holland: D. Reidel.

JEFFREYS, H. (1961) Theory of Probability. Oxford: Oxford University Press.

JOSHI, V. M. (1983) "Likelihood principle," pp. 644-647 in S. Kotz and N. J. Johnson (eds.) Encyclopedia of Statistical Sciences, Vol. 4. New York: John Wiley.

JUNGERMAN, H. (1983) "The two camps on rationality," pp. 63-86 in R. W. Scholz (ed.) Decision Making Under Uncertainty. Amsterdam, Holland: North-Holland.

KADANE, J. B. [ed.] (1976) "For what use are tests of hypotheses and tests of significance." (Symposium) Communications in Statistics—Theory and Methods A5: 735-787.

———J. M. DICKEY, R. L. WINKLER, W. S. SMITH, and S. C. PETERS (1980) "Interactive elicitation of opinion for a normal linear model." Journal of the American Statistical Association 75: 845-854.

KAHNEMAN, D., P. SLOVIC, and A. TVERSKY [eds.] (1982) Judgment Under Uncertainty: Heuristics and Biases. Cambridge: Cambridge University Press.

KEENEY, R. L. (1982) "Decision analysis: An overview." Operations Research 30: 803-838.

———and H. RAIFFA (1976) Decisions with Multiple Objectives. New York: John Wiley.

KEMPTHORNE, O. (1976) "Of what use are tests of significance and tests of hypotheses." Communications in Statistics—Theory and Methods A5: 763-777.

KINGMAN, J.F.C. (1975) "Review of theory of probability, a critical introductory treatment (vol. 1) by B. De Finetti." Journal of the Royal Statistical Society (A) 138: 98-99.

KRANTZ, D. H., R. D. LUCE, P. SUPPES, and A. TVERSKY (1971) Foundations of Measurement, Vol. 1. New York: Academic.

KRUSKAL, W. H. (1978) "Savage, Leonard Jimmie," pp. 889-891 in W. H. Kruskal and J. M. Tanur (eds.) International Encyclopedia of Statistics. New York: Free Press.

KYBURG, H. E., Jr. (1983) "Rational belief." The Behavioral and Brain Sciences 6: 231-273.

————(1974) The Logical Foundations of Statistical Inference. Dordrecht, Holland: D. Reidel.

————and H. E. SMOKLER [eds.] (1980a) Studies in Subjective Probability. Huntington, NY: Robert K. Krieger.

————(1980b) "Introduction," pp. 3-22 in H. E. Kyburg, Jr., and H. E. Smokler [eds.] Studies in Subjective Probability. Huntington, NY: Robert E. Krieger.

————[eds.] (1964) Studies in Subjective Probability. New York: John Wiley.

LAPLACE, P. S. (1951) Essays on Probabilities. New York: Dover. (orig. pub. 1820)

LARSON, H. J. (1974) Introduction to Probability Theory and Statistical Inference. New York: John Wiley.

LAU, J., J. P. KASSIRER, and S. G. PAUKER (1983) "Decision Maker 3.0: Improved decision analysis by personal computer." Medical Decision Making 3: 39-43.

LAUGHLIN, J. E. (1979) "A Bayesian alternative to least squares and equal weighting coefficients in regression." Psychometrika 44: 271-288.

LAVALLE, I. H. (1978) Fundamentals of Decision Analysis. New York: Holt, Rinehart & Winston.

————(1970) An Introduction to Probability, Decision, and Inference. New York: Holt, Rinehart & Winston.

LEAMER, E. E. (1978) Specification Searches: Ad Hoc Inference with Non-experimental Data. New York: John Wiley.

LEE, W. (1971) Decision Theory and Human Behavior. New York: John Wiley.

LEVI, I. (1982) "Direct inference and randomization." PSA 2: 447-463.

LEWIS, C. (1982) "Bayesian methods for the analysis of variance," pp. 283-306 in G. Keren (ed.) Statistical and Methodological Issues in Psychology and Social Sciences Research. Hillsdale, NJ: Lawrence Erlbaum.

LIBBY, D. L., M. R. NOVICK, J. J. CHEN, G. G. WOODWORTH, and R M. HAMER (1981) "The computer-assisted data analysis (CADA) monitor (1980)." American Statistician 35: 165-166.

LINDGREN, B. W. (1976) Statistical Theory. New York: Macmillan.

LINDLEY, D. V. (1982a) "Coherence," pp. 29-31 in S. Kotz and N. L. Johnson (eds.) Encyclopedia of Statistical Sciences 2. New York: John Wiley.

————(1982b) "The role of randomization in inference." PSA 2: 431-446.

————(1980) " L. J. Savage—his work in probability and statistics." Annals of Statistics 8: 1-24.

————(1978) "The Bayesian approach." Scandinavian Journal of Statistics 5: 1-26 (with discussion).

————(1976) "Bayesian statistics," pp. 353-362 in W. L. Harper and C. A. Hooker (eds.) Foundations of Probability Theory, Statistical Inference, and Statistical Theories of Science: Vol. 2. Foundations and Philosophy of Statistical Inference. Dordrecht, Holland: D. Reidel.

————(1972) Bayesian Statistics, A Review. Philadelphia: Society for Industrial and Applied Mathematics.

————(1971) Making Decisions. London: Wiley-Interscience.

————(1970) "Bayesian analysis in regression problems," pp. 37-56 in D. L. Meyer and R. O. Collier, Jr. (eds.) Bayesian Statistics. Itasca, IL: F. E. Peacock.

————(1965a) Introduction to Probability and Statistics from a Bayesian Viewpoint: Part 1, Probability. Cambridge: Cambridge University Press.

————(1965b) Introduction to Probability and Statistics from a Bayesian Viewpoint: Part 2, Inference. Cambridge: Cambridge University Press.

———(1957) "A statistical paradox." Biometrika 44: 187-192.

———and L. D. PHILLIPS (1976) "Inference for a Bernoulli process (a Bayesian view)." American Statistician 30: 112-119.

LUCE, R. D. and H. RAIFFA (1957) Games and Decisions: Introduction and Critical Survey. New York: John Wiley.

LUCE, R. D. and P. SUPPES (1965) "Preference, utility and subjective probability," pp. 249-410 in R. D. Luce et al. (eds.) Handbook of Mathematical Psychology. New York: John Wiley.

MacCRIMMON, K. R. and S. LARSSON (1979) "Utility theory: axioms versus 'paradoxes'," pp. 333-409 in M. Allais and O. Hagen (eds.) Expected Utility Hypotheses and the Allais Paradox: Contemporary Discussions of Decisions Under Uncertainty with Allais' Rejoinder. Dordrecht, Holland: D. Reidel.

MacKENZIE, D. A. (1981) Statistics in Britain 1865-1930: The Social Construction of Scientific Knowledge. Edinburgh: Edinburgh University Press.

McNEIL, B. J. and S. O. PAUKER (1984) "Decision analysis for public health: principles and illustrations," pp. 135-161 in L. Breslow et al. (eds.) Annual Review of Public Health, Vol. 5. Palo Alto, CA: Annual Reviews.

MEIER, P. (1981) "Jerome Cornfield and the methodology of clinical trials." Controlled Clinical Trials 1: 339-345.

MEYER, D. L. (1975) "Bayesian statistics." Presented at AERA National Convention, New York City, February 4-7, 1971. Reprinted in S. R. Houston, W. L. Duff, Jr, and R. M. Lynch (eds.) Applications in Bayesian Decision Processes. NY: MSS Information Corporation, pp. 5-9.

———(1964) "A Bayesian school superintendent." American Educational Research Journal 1: 219-228.

MOORE, P. G. (1978) "The mythical threat of Bayesianism." International Statistical Review 46: 67-73.

MORELL, J. A. (1979) Program Evaluation in Social Research. New York: Pergamon.

MORRISON, D. E. and R. E. HENKEL [eds.] (1970) The Significance Test Controversy. Chicago: Aldine.

MOSTELLER, F. and J. W. TUKEY (1968) "Data analysis, including statistics," pp. 80-203 in G. Lindsey and E. Aronson (eds.) The Handbook of Social Psychology, vol. 2. Reading, MA: Addison-Wesley.

MOSTELLER, F. and D. L. WALLACE (1964) Inference and Disputed Authorship: The Federalist. Reading, MA: Addison-Wesley.

———(1963) "Inference in an authorship problem." Journal of American Statistical Association 58: 275-309.

MURPHY, A. H. and R. L. WINKLER (1984) "Probability forecasting in meteorology." Journal of the American Statistical Association 79: 489-509.

NEYMAN, J. (1976) "Tests of statistical hypotheses and their use in studies of natural phenomena." Communications in Statistics—Theory and Methods A5: 737-751.

———(1952) Lectures and Conferences on Mathematical Statistics and Probability. Washington, DC: Department of Agriculture.

———(1950) First Course in Probability and Statistics. New York: Henry Holt.

———and E. S. PEARSON (1967a) "On the problem of the most efficient tests of statistical hypotheses," pp. 140-185 in Biometrika Trustees (eds.) Joint Statistical Papers: J. Neyman and E. S. Pearson. Berkeley: University of California Press. (reprinted from Philosophical Transactions of the Royal Society, London, Series A, 1933, 231: 289-337)

————(1967b) "The testing of statistical hypotheses in relation to probabilities *a priori*," pp. 186-202 in Biometrika Trustees (eds.) Joint Statistical Papers: J. Neyman and E. S. Pearson. Berkeley: University of California Press. (reprinted from proceedings of the Cambridge Philosophical Society, 1933, 24: 492-510).

NISBETT, R. and L. ROSS (1980) Human Inference: Strategies and Shortcomings of Social Judgment. Englewood Cliffs, NJ: Prentice-Hall.

NOVICK, M. R. (1980) "Statistics as psychometrics." Psychometrika 45: 411-424.

————and P. H. JACKSON (1974) Statistical Methods for Educational and Psychological Research. New York: McGraw-Hill.

NOVICK, M. R. and D. V. LINDLEY (1979) "Fixed-state assessment of utility functions." Journal of the American Statistical Association 74: 306-311.

————(1978) "The use of more realistic utility functions in educational applications." Journal of Educational Measurement 15: 181-191.

NOVICK, M. R., D. F. DEKEYREL, and D. T. CHUANG (1980a) "Local and regional coherence utility assessment procedures," pp. 557-568 in J. M. Bernardo et al. (eds.) Bayesian Statistics. Valencia, Spain: University Press.

NOVICK, M. R., R. M. HAMER, D. L. LIBBY, J. J. CHEN, and G. G. WOODWORTH (1980b) Manual for the Computer-Assisted Data Analysis (CADA) Monitor (1980). Iowa City: University of Iowa.

NOVICK, M. R., J. J. CHEN, S. MAYEKAWA, G. WOODWORTH, J. COOK, and R. L. NOVICK (1983) Manual for the Computer-Assisted Data Analysis (CADA) Monitor (1983). University of Iowa.

PEARSON, E. S. (1962) "Prepared contribution," pp. 53-58 in L. J. Savage et al., The Foundations of Statistical Inference. London: Methuen.

PHILLIPS, L. D. (1983) "A theoretical perspective on heuristics and biases in probabilistic thinking," pp. 525-543 in P. Humphreys et al. (eds.) Analysing and Aiding Decision Processes. Amsterdam, Holland: North-Holland.

————(1973) Bayesian Statistics for Social Scientists. New York: Thomas Y. Crowell.

PITZ, G. F. (1982) "Applications of Bayesian statistics in psychological research," in G. Keren (ed.) Statistical and Methodological Issues in Psychology and Social Sciences Research. Hillsdale, NJ: Lawrence Erlbaum.

————(1968) "An example of Bayesian hypothesis testing: the perception of rotary motion in depth." Psychological Bulletin 70: 252-255.

————and J. McKILLIP (1984) Decision Analysis for Program Evaluators. Beverly Hills, CA: Sage.

PITZ, G. F. and N. J. SACHS (1984) "Judgment and decision: theory and application," pp. 139-163 in M. R. Rosenzweig and L. W. Porter (eds.) Annual Review of Psychology, Vol. 35. Palo Alto, CA: Annual Reviews.

POLLARD, W. E. (1983) "Bayesian statistics and utilization of evaluation research findings: coherent inference and decision." Knowledge: Creation, Diffusion, Utilization 5: 56-83. (Corrigenda 5: 338)

PRATT, J. W. (1976) "A discussion of the question: for what use are tests of hypotheses and tests of significance." Communications in Statistics—Theory and Methods A5: 779-787.

————(1965) "Bayesian interpretation of standard inference statements." Journal of the Royal Statistical Society (B) 27: 169-203.

————(1961) "Review of testing statistical hypotheses by E. L. Lehmann." Journal of American Statistical Association 56: 163-167.

————, H. RAIFFA, and R. SCHLAIFER (1965) Introduction to Statistical Decision Theory (preliminary ed.) New York: McGraw-Hill.

————(1964) "The foundations of decision under uncertainty: an elementary exposition." Journal of American Statistical Association 59: 353-375.

PRESS, S. J. (1982) Applied Multivariate Analysis: Using Bayesian and Frequentist Methods of Inference. Malabar, FL: Robert E. Krieger.

————(1980) "Bayesian computer programs," pp. 429-442 in A. Zellner (ed.) Bayesian Analysis in Econometrics and Statistics: Essays in Honor of Harold Jeffreys. Amsterdam: North-Holland.

RAIFFA, H. (1968) Decision Analysis: Introductory Lectures on Choices Under Uncertainty. Reading, MA: Addison-Wesley.

————(1961) "Risk, ambiguity, and the Savage axioms: Comment." Quarterly Journal of Economics 75: 690-694.

————and R. SCHLAIFER (1961) Applied Statistical Decision Theory. Cambridge, MA: MIT Press.

RAMSEY, F. P. (1964) "Truth and probability," pp. 61-92 in H. E. Kyburg, Jr., and H. E. Smokler (eds.) Studies in Subjective Probability. New York: John Wiley. (pub. orig. in 1926)

RENYI, A. (1970) Foundations of Probability. San Francisco: Holden-Day.

RIECKEN, H. W. and R. F. BORUCH [eds.] (1974) Social Experimentation: A Method for Planning and Evaluating Social Intervention. New York: Academic.

ROBERTS, H. V. (1978) "Bayesian inference," pp. 9-16 in W. H. Kruskal and J. M. Tanur (eds.) International Encyclopedia of Statistics. New York: Free Press.

————(1977) "Reporting of Bayesian studies," pp. 155-173 in S. E. Fienberg and A. Zellner (eds.) Studies in Bayesian Econometrics and Statistics in Honor of Leonard Savage (vol. 2). Amsterdam: North-Holland.

————(1976) "For what use are tests of hypotheses and tests of significance." Communications in Statistics—Theory and Methods A5: 753-761.

————(1963) "Risk, ambiguity, and the Savage axioms: comment." Quarterly Journal of Economics 77: 327-342.

ROBINSON, G. K. (1982) "Behrens-Fisher problem," pp. 205-209 in S. Kotz and N. L. Johnson (eds.) Encyclopedia of Statistical Sciences (vol. 1). New York: John Wiley.

ROSSI, P. H. and H. E. FREEMAN (1982) Evaluation: A Systematic Approach. Beverly Hills, CA: Sage.

ROZEBOOM, W. W. (1960) "The fallacy of the null-hypothesis significance test." Psychological Bulletin 57: 416-428.

RUBIN, D. B. (1978) "Bayesian inference for causal effects: the role of randomization." Annals of Statistics 6: 34-58.

SAVAGE, L. J. (1976) "On rereading R. A. Fisher." Annals of Statistics 4: 441-500.

————(1972) The Foundations of Statistics. New York: Dover. (reprinted and revised from original 1954 edition, John Wiley.)

————(1964) "The foundations of statistics reconsidered," pp. 171-188 in H. E. Kyburg, Jr., and H. E. Smokler (eds.) Studies in Subjective Probability. New York: John Wiley. (reprinted and revised from Proceedings of the Fourth Berkeley Symposium on Mathematics and Probability, Vol. 1, 1961. Berkeley: University of California Press)

————(1962) "Subjective probability and statistical practice," pp. 9-35 (discussion pp. 62-103), in L. J. Savage (ed.) The Foundations of Statistical Inference. London: Methuen.

————(1961) The Subjective Basis of Statistical Practice. (Technical Report) Ann Arbor: Dept. of Statistics, University of Michigan. (not seen; described in Berger, 1980)

SCHLAIFER, R. (1971) Computer Programs for Elementary Decision Analysis. Cambridge, MA: Division of Research, Graduate School of Business Administration, Harvard University.

————(1969) Analysis of Decisions Under Uncertainty. New York: McGraw-Hill.

————(1961) Introduction to Statistics for Business Decisions. New York: McGraw-Hill.

————(1959) Probability and Statistics for Business Decisions. New York: McGraw-Hill.

SCHMITT, S. A. (1969) Measuring Uncertainty: An Elementary Introduction to Bayesian Statistics. Reading, MA: Addison-Wesley.

SCHOLZ, R. W. [ed.] (1983) Decision Making Under Uncertainty: Cognitive Decision Research, Social Interaction, Development and Epistemology. Amsterdam, Holland: North-Holland.

SEAL, H. L. (1978) "Bayes, Thomas," pp. 7-9 in W. H. Kruskal and J. M. Tanur (eds.) International Encyclopedia of Statistics. New York: Free Press.

SEIDENFELD, T. (1979) Philosophical Problems of Statistical Inference: Learning from R. A. Fisher. Dordrecht, Holland: D. Reidel.

SILVEY, S. D. (1975) Statistical Inference. London: Chapman and Hall.

SJÖBERG, L., T. TYSZKA, and J. A. WISE [eds.] (1983) Human Decision Making. Bodafors, Sweden: Doxa.

SLOVIC, P. and A. TVERSKY (1974) "Who accepts Savage's axiom?" Behavioral Science 19: 368-373.

SLOVIC, P., B. FISCHHOFF, and S. LICHTENSTEIN (1977) "Behavioral decision theory," pp. 1-39 in M. R. Rosenzweig and L. W. Porter (eds.) Annual Review of Psychology, Vol. 28. Palo Alto, CA: Annual Reviews.

SMITH, A.F.M. (1984) "Present position and potential developments: some personal views—Bayesian statistics." Journal of the Royal Statistical Society (A) 147(Part 2): 245-259.

SMITH, C.A.B. (1965) "Personal probability and statistical analysis." Journal of the Royal Statistical Society (A) 128: 469-499.

SPROTT, D. A. and J. G. KALBFLEISCH (1965) "Use of the likelihood function in inference." Psychological Bulletin 64: 15-22.

SUPPES, P. (1982) "Arguments for randomizing." PSA 2: 464-475.

SUSARLA, V. (1982) "Empirical Bayes theory," pp. 490-503 in S. Kotz and N. L. Johnson (eds.) Encyclopedia of Statistical Sciences (vol. 2). New York: John Wiley.

THOMPSON, M. S. (1982) Decision Analysis for Program Evaluation. Cambridge, MA: Ballinger.

TUKEY, J. W. (1960) "Conclusions vs. decisions." Technometrics 2: 423-433.

TVERSKY, A. and D. KAHNEMAN (1981) "The framing of decisions and the psychology of choice." Science 211: 453-458.

————(1974) "Judgment under uncertainty: heuristics and biases." Science 185: 1124-1131.

VON NEUMANN, J. and O. MORGENSTERN (1947) Theory of Games and Economic Behavior. Princeton, NJ: Princeton University Press.

WALD, A. (1950) Statistical Decision Functions. New York: Chelsea.

WALLSTEN, T. S. [ed.] (1980) Cognitive Processes in Choice and Decision Behavior. Hillsdale, NJ: Lawrence Erlbaum.

————and D. V. BUDESCU (1983) "Encoding subjective probabilities: a psychological and psychometric review." Management Science 29: 151-173.

WANG, M., M. R. NOVICK, G. L. ISAACS, and D. OZENE (1977) "A Bayesian data analysis system for the evaluation of social programs." Journal of American Statistical Association 72: 711-722.

WEINSTEIN, M. C. and H. V. FINEBERG (1980) Clinical Decision Analysis. Philadelphia: W. B. Saunders.

WINER, B. J. (1971) Statistical Principles in Experimental Design. New York: McGraw-Hill.

———(1977) "Prior distributions and model building in regression analysis," pp. 233-242 in A. Aykaç and C. Brumat (eds.) New Developments in the Application of Bayesian Methods. Amsterdam, Holland: North-Holland.

WINKLER, R. L. (1983) "Judgments under uncertainty," pp. 332-336 in S. Kotz and N. L. Johnson (eds.) Encyclopedia of Statistical Sciences (vol. 4). New York: John Wiley.

———(1982a) "The Bayesian approach: a general review," pp. 217-244 in G. Keren (ed.) Statistical and Methodological Issues in Psychology and Social Sciences Research. Hillsdale, NJ: Lawrence Erlbaum.

———(1982b) "Research directions in decision making under uncertainty." Decision Sciences 13: 517-533.

———(1972) Introduction to Bayesian Inference and Decision. New York: Holt, Rinehart & Winston.

WRIGHT, G. (1984) Behavioral Decision Theory. Beverly Hills, CA: Sage.

ZELLNER, A. (1971a) An Introduction to Bayesian Inference in Econometrics. New York: John Wiley.

———(1971b) "The Bayesian approach and alternatives in econometrics—I," pp. 178-193 in M. D. Intriligator (ed.) Frontiers of Quantitative Economics. Amsterdam: North-Holland.

———and A. SIOW (1980) "Posterior odds ratios for selected regression hypotheses," pp. 585-604 in J. M. Bernardo et al. (eds.) Bayesian Statistics. Valencia, Spain: Valencia University Press.

Index

Aitchison, J., 233, 241
Allais, M., 55, 58, 241
American Statistical Association and the Institute of Mathematical Statistics, 22, 241
Anscombe, F. J., 121, 241

Bakan, D., 109, 241
Barnard, G. A., 17, 241
Barnes, V., 241
Barnett, V., 22, 41-42, 46, 54, 121, 207, 241
Bayes, T., 17, 18, 22, 46, 241
Bayes' postulate, 18
Bayes' theorem, 13, 18, 45-46, 59-76; likelihood, 63; odds-likelihood ratio form, 63; posterior probability, 63; prior probability, 63
Bayesian inference, 77-208; comparing means, 102-118, 146-158, 172-173; comparing variances, 146-158; correlation coefficients, 174-175; proportions, 188-208, 200-202; regression parameters, 159-187; single means, 78-87, 100-102, 123-146; single variances, 123-146
Beach, L. R., 56, 58, 241-242
Behn, R. D., 233, 241
Berger, J. O., 58, 89-90, 112, 121, 122, 240-241
Berkeley, D., 55-56, 241
Bernardo, J. M., 76, 121, 238, 242

Birnbaum, A., 208, 242
Boruch, R. F., 8, 249
Box, G.E.P., 76, 90, 95-97, 99-100, 121, 127, 129, 242
Break-even value, 217-218
Broemling, L. D., 187, 242
Brown, R. V., 211, 233, 242
Budescu, D. V., 58, 251

Chen, J. J., 233, 242, 246, 248
Chernoff, H., 233, 242
Christensen-Szalanski, J., 56, 241-242
Chuang, D. T., 248
Classical inference, 23-42; comparing means, 32, 38, 106-114, 155-156; comparing variances, 32-33; proportions, 197-198; regression parameters, 163-165, 179, 185-187; single means, 30-31, 37-38, 147-148; single variances, 32-33
Classical methods, 23-42; interval estimation, 37-39, 112-113; hypothesis testing, 25-37, 106-112, 118-121; point estimation, 39-40, 113-114; significance testing, 26-28
Cohen, L. J., 58, 242
Coherence, axioms, 51-57
Computer Assisted Data Analysis (CADA) Monitor, 124
Conjugate distributions, 82-83

Cook, J., 248
Cornfield, J., 121, 202, 206, 242
Cox, D. R., 40-41, 242
Crane, J. A., 22, 242
Credible interval, 113

Decision making, 51-59, 209-234; Bayesian, 51-58, 209-234; classical, 212-223
Deely, J. J., 8, 242
De Finetti, B., 18-20, 22, 49, 54, 91, 242, 243
DeGroot, M. H., 58, 121, 233, 242-243
DeKeyrel, D. F., 245, 248
Dickey, J., 97, 243, 245

Edwards, A.W.F., 204, 208, 243
Edwards, W., 22, 49, 52, 56, 95, 98, 108, 121, 214, 233, 243
Eells, E., 58, 243
Einhorn, H. J., 58, 243
Ellsberg, D., 58, 243
Enthoven, A. C., 225, 243
Equivalent prior sample, 85-87
Error characteristic curve, 33-35

Fairweather, G. W., 8, 243
Farquhar, P. H., 233, 243
Feather, N. T., 58, 243
Fellner, W., 58, 243
Fennessey, J., 22, 244
Ferguson, T. S., 233, 244
Fine, T. L., 58, 244
Fineberg, H., 76, 251
Fischhoff, B., 233, 244, 250
Fishburn, P. C., 51, 58, 244
Fisher, R. A., 9, 19, 22, 26-28, 41, 155-156, 158, 208, 244
Foundations of probability and statistics, 42, 244
Freeling, A.N.S., 233, 244
Freeman, H. E., 8, 249

Geisser, S., 238, 244
Gokhale, D. V., 175, 244
Good, I. J., 21, 41, 57, 207, 238, 244
Greenwald, A. G., 121, 244
Guttentag, M., 243

Hagen, O., 58, 241
Hamaker, H. C., 20, 244

Hamer, R. M., 246, 248
Hampton, J. M., 121, 244
Hartigan, J. A., 57, 244
Hays, W. L., 41, 121, 157, 244
Henkel, R. E., 42, 247
Highest Density Region (HDR), 112-113
Hildreth, C., 98, 245
Hinkley, D. V., 40-41, 242
Hogarth, R. M., 58, 243, 245
Holland, J. D., 17, 22, 245
Houle, A., 11, 245
Howard, R. A., 233, 245
Huber, G. P., 225, 233, 245
Hull, J., 233, 245
Humphreys, P., 55-56, 83, 241, 245
Hunter, D. E., 76, 245

Isaacs, G. L., 251
Iverson, G., 121, 245

Jackson, P. H., 11, 22, 57, 76, 91, 97, 99, 113, 121, 127, 137, 155, 157-158, 161, 171, 175, 187, 207, 236, 245, 248
Jaynes, E. T., 121, 245
Jeffreys, H., 19, 48, 118, 121, 202, 205, 245
Joshi, V. M., 208, 245
Jungerman, H., 55, 245

Kadane, J. B., 42, 187, 245
Kahneman, D., 55, 58, 245, 251
Kahr, A. S., 242
Kalbfleisch, J. B., 208, 250
Kassirer, J. P., 246
Keeney, R. L., 210, 233, 245
Kempthorne, O, 121, 245
Keynes, J. M., 48
Kingman, J.F.C., 49, 245
Krantz, D. H., 58, 246
Kruskal, W. H., 22, 246
Kyburg, H. E., Jr., 10, 19, 28, 42, 57-58, 246

Laplace, P. S., 18, 46, 246
Larson, H. J., 41, 246
Larsson, S., 58, 247
Lau, J., 231, 246
Laughlin, J. E., 177, 246
LaValle, I. H., 52, 58, 233
Leamer, E. E., 12, 48, 88, 187, 239, 246
Lee, W., 58, 246
Levi, I., 122, 246

Lewis, C., 158, 246
Libby, D. L., 124, 246, 248
Lichtenstein, S., 250
Likelihood function, 75, 79-80, 124-125, 161-162, 189-190, 200-208
Likelihood principle, 202, 208
Lindgren, B., 41, 99, 233
Lindley, D. V., 8, 20, 22, 53-54, 57-58, 118, 121-122, 157-158, 161, 174, 187, 207, 233, 242, 246-248
Lindley's Paradox, 118-121
Lindman, H., 243
Loss function, 216-218
Luce, R. D., 58, 215, 246-247

MacCrimmon, K. R., 58, 247
MacKenzie, D. A., 22, 247
McKillip, J., 233, 248
McNeil, B. J., 233, 247
Meier, P., 88, 247
Meyer, D. L., 109, 121, 247
Moore, P. G., 21, 244-245, 247
Morell, J. A., 8, 247
Morgenstern, O., 20, 251
Morrison, D. E., 42, 247
Moses, L. E., 233, 242
Mosteller, R., 20, 92, 98-99, 247
Murphy, A. H., 56, 247

Newman, J. R., 233, 243
Neyman, J., 9, 19, 27-28, 33, 36, 41-42, 109, 121, 219, 222, 248
Nisbett, R., 55, 248
Novick, M. R., 11-12, 22, 57, 76, 91, 99, 113, 121, 124, 127, 137, 155, 157-158, 161, 171, 175, 187, 207, 233, 242, 245-246, 248, 251
Novick, R. L., 248

Ozene, D., 251

Pauker, S. G., 233, 246-247
Pearson, E. S., 9, 19, 27-28, 33, 36, 109, 219, 222, 248
Peters, S. C., 245
Peterson, C., 242
Phillips, L. D., 56, 76, 113, 121, 156-158, 187, 199, 207, 243, 247-248
Pitz, G. F., 58, 121, 137, 158, 199, 207, 233, 248

Pollard, W. E., 8, 22, 249
Pratt, J. W., 28, 39, 48, 92, 121, 187, 207, 233-234, 249
Press, S. J., 158, 175, 244, 249
Precision, initial and final, 40-41
Precision of a normal distribution, 84-85
Prior distributions, 87-102; assessment, 91-96; data-based, 91-92; non-data-based, 92-94; noninformative, 94-96
Probability, 43-58; classical interpretation, 46-47; conditional, 45; conditional independence, 66-67; frequency interpretation, 46-48; independence, 45; Kolmogorov axioms, 44; logical interpretation, 48-49; posterior, 63; prior, 63; subjective interpretation, 49-50
Probability distributions, 68; Behrens, 155; Bernoulli, 189; beta, 190-192, 196-199; binomial, 189-190, 196-198; chi-square, 32-33; conditional, 70-73; F, 33, 156-157; inverse chi, 127-128; inverse chi-square, 127-128; joint, 70-73; log-uniform, 125-126; marginal, 70-73; multivariate normal, 178; multivariate t, 178, 179; normal, 30-32, 78-80, 124-125; t, 31-32, 128-129; uniform, 94-96

Raiffa, H., 12, 20, 22, 58, 80, 82, 121, 187, 207, 210, 215, 221, 227, 233-234, 236, 245, 247, 249
Ramsey, F. P., 19-20, 49, 249
Reicken, H. W., 8, 249
Renyi, A., 57, 249
Risk function, 219
Roberts, H. V., 58, 97, 100, 121-122, 249
Robinson, G. K., 156, 249
Ross, L., 55, 248
Rossi, P. H., 8, 249
Rozeboom, W. W., 109, 249
Rubin, D. B., 122, 249

Sachs, N. J., 58, 248
Savage, L. J., 20, 22, 41, 49, 58, 90, 95, 207, 243, 250
Schlaifer, R., 12, 20, 22, 80, 82, 121, 158, 187, 207, 223, 227, 233-234, 236, 249-250
Schmitt, S. A., 76, 157-158, 187, 207, 250
Scholz, R. W., 58, 250
Seal, H. L., 22, 250
Seidenfeld, T., 41, 250

Silvey, S. D., 36, 41, 250
Siow, A., 186, 187
Sjöberg, L. T., 58, 250
Slovic, P., 55, 58, 245, 250
Smith, A.F.M., 54, 242, 250
Smith, C.A.B., 121, 250
Smith, W. S., 245
Smokler, H. E., 57-58, 246
Snapper, K., 243
Sprott, D. A., 208, 250
Stopping rules, 200-207; binomial sampling, 200-202; Pascal sampling, 200-202
Sufficient statistics, 80
Suppes, P., 58, 122, 246-247, 250
Susarla, V., 8, 250
Svenson, O., 241, 245

Thomas, H., 244-245
Thompson, M. S., 233, 250
Tiao, G. C., 95-97, 99-100, 121, 127, 129, 242
Tornatzky, L. G., 8, 243
Tukey, J. W., 41, 92, 247, 250
Tversky, A., 55, 58, 245-246, 250-251
Tyszka, T., 250

Utility, 53-54, 212-214, 216-217
Utility, expected, 53

Vari, A., 245
Vaupel, J. W., 233
Von Neumann, J., 19-20, 251

Wagenaar, W. A., 241
Wald, A., 19, 109, 219, 222-223, 251
Wallace, D. L., 20, 98-99, 247
Wallsten, T. S., 58, 251
Wang, M., 22, 251
Weinstein, M. C., 76, 251
Winer, B. J., 156, 251
Winkler, R. L., 48, 56, 58, 76, 89, 121, 187, 207, 227, 233, 244-245, 247, 251
Wise, J. A., 250
Woodworth, G. G., 246, 248
Wright, G., 58, 251

Yates, F., 155, 244

Zellner, A., 40, 92, 109, 114, 121, 163, 186-187, 251

About the Author

William E. Pollard is Assistant Professor in the Department of Psychiatry at Emory University School of Medicine and is Director of Research, Evaluation, and Planning for the Psychiatry Department at Grady Memorial Hospital and Central Fulton Community Mental Health Center in Atlanta, Georgia. He holds a B.A. degree in psychology from the University of Minnesota and a Ph.D. degree in psychology from the University of Washington. He was a NIMH postdoctoral fellow in the Methodology and Evaluation Program in the Department of Psychology at Northwestern University. His interests are in quantitative methods for evaluation and decision making. He has published articles and presented papers in the areas of health status measurement, research utilization, Bayesian statistics, computer security, computerized information systems, and management of research and software development projects.

AUNR